Liberty Hyde Bailey

Cross-breeding and hybridization

The philosophy of the crossing of plants, considered with reference to

their improvement under cultivation

Liberty Hyde Bailey

Cross-breeding and hybridization
The philosophy of the crossing of plants, considered with reference to their improvement under cultivation

ISBN/EAN: 9783337146917

Printed in Europe, USA, Canada, Australia, Japan

Cover: Foto ©berggeist007 / pixelio.de

More available books at **www.hansebooks.com**

A Lecture given before the Massachusetts State Board of Agriculture in Boston, Dec. 1, 1891.

CROSS-BREEDING AND HYBRIDIZING

THE PHILOSOPHY OF THE CROSSING OF PLANTS, CONSIDERED WITH REFERENCE TO THEIR IMPROVEMENT UNDER CULTIVATION; WITH A BRIEF BIBLIOGRAPHY OF THE SUBJECT.

By L. H. BAILEY.

NEW YORK:
THE RURAL PUBLISHING COMPANY.
1892.

ELECTROTYPED AND PRINTED BY
THE RURAL PUBLISHING COMPANY.

Cross-Breeding and Hybridizing.

THE PHILOSOPHY OF THE CROSSING OF PLANTS, CONSIDERED WITH REFERENCE TO THEIR IMPROVEMENT UNDER CULTIVATION.

IT is now understood that the specific forms or groups of plants have been determined largely by the survival of the fittest in a long and severe struggle for existence. The proof that this struggle everywhere exists becomes evident upon a moment's reflection. We know that all organisms are eminently variable. In fact, no two plants or animals in the world are exactly alike. We also know that very few of the whole number of seeds which are produced in any area ever grow into plants. If all the seeds produced by the elms upon Boston Common in any fruitful year were to grow into trees, this city would become a forest as a result. If all the seeds of arest orchid in our woods were to grow, in a few generations of plants our farms would be overrun. If all the rabbits which are born were ᐢh old age and all their offspring were to do the same, in less than rs every vestige of herbage would be swept from the country and our would become barren. There is, then, a wonderful latent potency se species; but the same may be said of every species of plant and ᑊ, even of man himself. If one species of plant would overrun and ᐧe land if it increased to the full extent of its possibilities, what ᑊ ᵹ the result if each of the 2,061 plants known to inhabit Middlesex were to do the same? And then fancy the result if each of the ani-ᔆ, from rabbits and mice to frogs and leeches, were to increase without ᵏk! The plagues of Egypt would be insignificant in the comparison! fact is, the world is not big enough to hold the possible first offspring of ᐧlants and animals at this moment living upon it. Struggle for existence, then, is inevitable, and it must be severe. It follows as a necessity that those seeds grow or those plants live which are best fitted to grow and to live, or which are fortunate enough to find a congenial foothold. It would appear at first thought that much depends upon the accident of falling into a congenial place, or one unoccupied by other plants or animals; but inasmuch as scores of plants are contending for every unoccupied place, it follows that everywhere only the fittest can germinate or grow. In the great majority of cases, plants grow in a certain place because they are better fitted to grow there, to hold their own, than any other plants are; and the instances are rare in which a plant is so fortunate as to find an unoccupied place. We are apt to think that plants chance to grow where we find them, but the chance is determined by law and therefore is not chance!

Much of the capability of a plant to persist under all this struggle, therefore, depends upon how much it varies; for the more it varies the more likely it is to find places of least struggle. It grows under various conditions—in sun and shade, in sand and clay, by the seashore or upon the hills, in the humidity of the forest or the aridity of the plain. In some directions it very likely finds less struggle than in others, and in these directions it expands itself, multiplies, and gradually dies out in other directions. So it happens that it tends to take on new forms or to undergo an evolution. In the meantime, all the intermediate forms, which are at best only indifferently adapted to their conditions, tend to disappear. In other words, gaps appear which we call "missing links." The weak links break and fall away, and what was once a chain becomes a series of rings. So the "missing links" are among the best proofs of evolution.

The question now arises as to the cause of these numerous variat' animals and plants. Why are no two individuals in nature exactly alike; ne question is exceedingly difficult to answer. It was once said that plants v because it is their nature to vary, that variation is a necessary function, much as growth or fructification. This really removes the question yond the reach of philosophy; and direct observation leads us to think some variation, at least, is due to external circumstances. We are now l ing for the cause of variation along some of the lines of evolution, an are wondering if the varied surroundings, or, as Darwin put it, the "cha conditions of life," may not actually induce variability. This cor would seem to follow to some extent from the fact of the severe a versal struggle in nature whereby plants are constantly forced into new strange conditions. But there is undoubtedly much variation whi sprung from more remote causes, one of which it is my purpose to di here.

In the lowest animals and plants the species multiplies by mean simple division, or by budding.   One individual, of itself, becomes two, and the two are therefore recasts of the one.   But as organisms multiplied and conditions became more complex, that is, as struggle increased, there came a differentiation in the parts of the individual, so that one cell or one cluster of cells performed one labor and other cells performed other labor ; and this tendency resulted in the development of organs.   Simple division, therefore, could no longer reproduce the whole complex individual, and as all organs are necessary to the existence of life, the organism dies if it is divided. Along with this specialization came the differentiation into sex, and sex clearly has two offices—to hand over, by some mysterious process, the complex organization of the parent to the offspring, and also to unite the essential characters or tendencies of two beings into one.   The second office is manifestly the greater, for, as it unites two organizations into one, it insures that the offspring is somewhat unlike either parent and is, therefore, better fitted to seize upon any place or condition new to its kind.   And as the generations increase, the tendency to variation in the offspring must be constantly greater, because the impressions of a greater number of ancestors are transmitted to it.   I have said that this office of sex to induce variation is more important than the mere fact of reproduction of a complex organization, for it must be borne in mind that the complexity of organization is itself a variation made necessary by the increasing struggle for existence.

If, therefore, the philosophy of sex is to promote variation by the union of different individuals, it must follow that greatest variation must come from parents considerably unlike each other in their minor characters. Thus it comes that inbreeding tends to weaken a type and cross-breeding tends to strengthen it.   And at this point we meet the particular subject which I am to present to you.   I have introduced to you this preliminary sketch because I contend that we can understand crossing only as we make it a part of the general philosophy of nature.   There are the vaguest notions concerning the possibilities of crossing, some of which I hope to correct by presenting this subject in its relations to the general aspects of the vegetable world.

We are now prepared to understand that crossing is good for the species, because it constantly revitalizes offspring with the strongest traits of the parents and ever presents new combinations which enable the individuals to stand a better chance of securing a place in the polity of nature.   All the further discussions of the subject are such as have to do with the extent to which crossing is possible and advisable, and the mere methods of performing the operation.

At this point I must digress for the purpose of defining certain terms which it is necessary to use frequently.   I use the term *cross* to denote the offspring of any sexual union between plants, whether of different species, or varieties, or even different flowers upon the same plant. It is a general term. And the word is also sometimes used to denote the operation of performing or bringing about the sexual union. There are different kinds of crosses. One of these is the *hybrid.* A hybrid is a cross between two species, as a plum and a peach, or a raspberry and a blackberry. There has lately been some objection urged against this term because it is often impossible to define the limitations of species, to tell where one species ends and another begins. And it is a fact that this difficulty exists, for plants which some botanists regard as mere varieties others regard as distinct species. But the term hybrid is no more inaccurate than the term species, upon which it rests, and so long as men talk about species, so long have we an equal right to talk about hybrids. Here, as everywhere, terms are mere conveniences, and they seldom express the whole truth. In common speech, the word hybrid is much misused. Crosses between varieties of one species are termed *half-breeds* or *cross-breeds*, and those between different flowers upon the same plant are called *individual crosses.*

If crossing is good for the species, which philosophy and direct experiment abundantly show, it is necessary at once to find out to what extent it can be carried. Does the good increase in proportion as the cross becomes more violent, or as the parents are more and more unlike? Or do we soon find a limit beyond which it is not profitable or even possible to go—a point at which we say that "An inch is as good as an ell"? If great variability is good for the species in the struggle for existence, and if crossing induces variability because of the union of unlike individuals, it would seem to follow that the more unlike the parents are the greater would be the variation in offspring and the more the species would prosper; and carrying this thought to its logical conclusion we should expect to find that the most closely related plants would constantly tend to refuse to cross, because the offspring of them would be little variable and, therefore, little adapted to the struggle for existence, while the most widely separated plants would constantly tend to cross more and more, because their offspring would present the greatest possible degrees of differences. We should expect, for instance, that a Baldwin apple would be less likely to cross with a Greening than with a Norway spruce or Indian corn! And if we should carry our thought a step farther, we should at once see that this crossing between different species would soon fill in all differences between those species, and that definite specific types would cease to exist. This would be pandemonium, and crossing would be the cause of it!

Now, essentially this reasoning has been advanced to combat the evolution of plants and animals by means of natural selection, and this proposition that intermixing must constantly tend to obliterate all differences between plants and to prevent the establishment of well-marked types, has been called the "swamping effects of intercrossing." It is exceedingly important that we consider this question, for it really lies at the foundation of the improvement of cultivated plants by means of crossing, as well as of the persistence and evolution of varieties and species under wholly natural conditions.

We find, however, that distinct species, as a rule, refuse to cross, and the first question which naturally arises in this discussion is, What is the immediate cause of this refusal of plants to cross? How does this refusal express itself? It comes about in many ways. The commonest cause is the positive refusal of a plant to allow its ovules to be impregnated by the pollen of another plant. The pollen will not "take." For instance, if we apply the pollen of a Hubbard squash to the flower of the common field pumpkin there will simply be no result—the fruit will not form. The same is true of the pear and the apple, the oat and the wheat, and most very unlike species. Or the refusal may come in the sterility of the cross or hybrid. The pollen may "take" and seeds may be formed and the seeds may grow, but the plants which they produce may be wholly barren, sometimes even refusing to produce flowers as well as seeds, as in the instance of some hybrids between the Wild Goose plum and the peach. Sometimes the refusal to cross is due to some difference in the time of blooming or some incompatability in the structure of the flowers. But it is enough for our purpose to know that there are certain characters in widely dissimilar plants which prevent intercrossing, and that these characters are just as positive and just as much influenced by change of environment and natural selection, as are size, color, productiveness and other characters. Here, then, is the sufficient answer to the proposition that intercrossing must swamp all natural selection, and also the explanation of the varying and often restricted limits within which crossing is possible : that is, the checks to crossing have been developed through the principle of universal variability and natural selection, as has been shown by Darwin and Wallace. Plants vary in their reproductive organs and powers just the same as they do in other directions, and when such a variation is useful it is perpetuated, and when hurtful it is lost. Suppose that a certain well-marked individual of a species should find an unusually good place in nature and it should multiply rapidly ; crosses would be made between its own offspring and perhaps between those offspring and itself in succeeding years, and it is fair to suppose that some of the crosses would be particularly well-adapted to the conditions in which the parent grew, and these would constantly tend to perpetuate themselves, while less adaptive forms would constantly tend to disappear. Now the same thing

Would take place if this individual or its adaptive offspring were to cross with the main stock of the parent species; for all the offspring of such a cross which are intermediate in character and therefore less adapted to the new conditions, would tend to disappear, and the two branches of the species would, as a result, become more and more fixed and the tendency to cross would constantly decrease. The refusal to cross, therefore, becomes a positive character of separation, and the "missing links" which result from crossing are no more and no less inexplicable than the "missing links" due to simple selection; or, to put the case more accurately, natural selection weeds out the tendency to promiscuous crossing when it is hurtful, in just the same manner that it it weeds out any other injurious tendency. It makes no difference in what way this tendency expresses itself, whether in some constitutional refusal to cross—if such exists—or in infertility of offspring, or in different times of blooming: all equally come under the same power of natural selection. We are apt to look upon infertility as the absence of a character, a sort of a negative feature, which is somehow not the legitimate property of natural selection; but such is not the case. We are perhaps led the more to this feeling because the word infertility is itself negative, and because we associate full productiveness with the positive attributes of plants. But loss of productiveness is surely no more a subject of wonder than loss of color or size, if there is some corresponding gain to be accomplished. In fact, we see in numerous plants which propagate easily by means of runners and suckers, a very low degree of productiveness.

Now, if this reasoning is sound, it leads us to conclusions quite the reverse of those held by the advocates of the swamping effects of intercrossing, and these conclusions are of the most vital importance to every man who tills the soil. The logical result is simply this: the best results of crossing are obtained, as a rule, when the cross is made between different individuals of the same variety, or, at farthest, between different individuals of the same species. In other words, hybrids—or crosses between species—are rarely useful; and it follows, as a logical result, that the more unlike the species, the less useful will be the hybrids. This, I am aware, is counter to the notions of most horticulturists, and, if true, must entirely overthrow our common thinking upon this subject. But I think that I shall be able to show that observation and experiment lead to the same conclusion to which our philosophy has brought us.

At this point we must ask ourselves what we mean by "best results." I take this phrase to refer to those plants which are best fitted to survive in the struggle for existence—those which are most vigorous or most productive or most hardy, or which possess any well-marked character or characters which distinguish them in virility from their fellows. We commonly associate the term more particularly with marked vigor and productiveness; these are the characters most useful in nature and also in cultivation, the ones which we oftenest desire to obtain. Another type of variation which we constantly covet is something which we can call a new character, which will lead to the production of a new cultural variety; and we are always looking to this as the legitimate result of crossing. We have forgotten, if, indeed, we ever knew, that the commoner, all-pervading, more important function of the cross is to infuse some new strength or power into the off-spring—to improve or to perpetuate an existing variety rather that to create a new one. Or, if a new one is created, it comes from the gradual passing of one into another—an inferior variety into a good one, a good one into a superlative one. So nature employs crossing in a process of slow or gradual improvement, one step leading to another, and not in any bold or sudden creation of new forms. And there is evidence to show that something akin to this must be done to secure the best and most permanent results under cultivation. The notion is somehow firmly rooted in the popular mind that new varieties can be produced with the greatest ease by crossing parents of given attributes. There is something captivating about the notion. It smacks of a somewhat magic power which man evokes as he passes his wand over the untamed forces of nature. But the wand is often only a gilded stick, and is apt to serve no better purpose than the drum-major's pretentious baton!

Let me say, further, that crossing alone can accomplish comparatively little. The chief power in the evolution or progression of plants appears to be selection, or, as Darwin puts it, the law of "preservation of favorable individual differences and variations, and the destruction of those which are injurious." Selection is the force which augments, develops and fixes types. Man must not only practice a judicious selection of parents from which the cross is to come, which is in reality but the exercise of a choice, but he must constantly select the best from among the crosses in order to maintain a high degree of usefulness and to make any advancement; and it sometimes happens that the selection is much more important to the cultivator than the crossing. I do not wish to discourage the crossing of plants, but I do desire to dispell the charm which too often hangs about it.

Further discussion of this subject naturally falls under two heads: the improvement of existing types or varieties by means of crossing, and the summary production of new varieties. I have already stated that the former office is the more important one, and the proposition is easy of proof. It is the chief use which nature makes of crossing—to strengthen the type. Think, for instance, of the great rarity of hybrids or pronounced crosses in nature. No doubt all the authentic cases on record could be entered in one or two volumes ; but a list of all the individual plants of the world could not be compressed into ten thousand volumes. There are a few genera in which the species are not well defined, or in which some character of inflorescence favors promiscuous crossing, in which hybrids are conspicious ; but even here the number of individual hybrids is very small in comparison with the whole number of individuals. That is, the hybrids are rare, while the parents may be common. This is well illustrated even in the willows and oaks, in which, perhaps, hybrids are better known than in any other American plants. The great genus carex or sedge, which occurs in great numbers and many species in almost every locality in New England, and in which the species are particularly adapted to intercrossing by the character of their inflorescence, furnishes but few undoubted hybrids. Among 167 species and prominent varieties inhabiting the northeastern states, there are only nine hybrids recorded, and all of them are rare or local, some of them having been collected but once. Species of remarkable similarity may grow side by side for years, even intertangled in the same clump, and yet produce no hybrid. These instances prove that nature avoids hybridization, a conclusion at which we have already arrived from philosophical considerations. And we have reason to infer the same conclusion from the fact that flowers from different species are so constructed as not to invite intercrossing. But, on the other hand, the fact that all higher plants habitually propagate by means of seeds, which is far the most expensive to the plant of all methods of propagation, while at the same time most flowers are so constructed as to prevent self-fertilization, prove that some corresponding good must come from crossing within the limits of the species or variety ; and there are purely philosophical reasons, as we have seen, which warrant a similar conclusion. But experiment has given us more direct proof of our proposition, and we shall now turn our attention to the garden.

Darwin was the first to show that crossing within the limits of the species or variety results in a constant revitalizing of the offspring, and that this is the particular ultimate function of cross-fertilization. Kölreuter, Sprengel, Knight, and others, had observed many, if, indeed, not all the facts obtained by Darwin, but they had not generalized upon them broadly and did not conceive their relation to the complex life of the vegetable world. Darwin's results are, concisely, these : Self-fertilization tends to weaken the offspring, as compared with its natural condition ; crossing between different plants of the same variety gives stronger and more productive offspring than arises from self-fertilization ; crossing between stocks of the same variety grown in different places or under different conditions gives better offspring than crossing between different plants grown in the same place or under similar conditions ; and his researches have also shown that, as a rule, flowers are so constructed as to favor cross-fertilization. In short, he found, as he expressed it, that " nature abhors perpetual self-fertilization." Some of his particular results, although often quoted, will be useful in fixing these facts in our minds. Plants from crossed seeds of morning-glory exceeded in height those from self-fertilized seeds as 100 exceeds 76 in the first generation. Some flowers upon these plants were self-pollinated and some were crossed, and in this second generation the crossed plants were to the uncrossed as 100 is to 79 ; the operation was again repeated, and in the third generation the figures stand 100 to 68 ; fourth generation, the plants having been grown in midwinter when none of them did well, 100 to 86 ; fifth generation, 100 to 75 ; sixth generation, 100 to 72 ; seventh generation, 100 to 81 ; eighth generation, 100 to 85 ; ninth generation, 100 to 79 ; tenth generation, 100 to 54. The average total gain in height of the crossed over the uncrossed was as 100 to 77, or about 30 per cent. There was a corresponding gain in fertility, or the number of seeds and seed-pods produced. Yet, striking as these results are, they were produced by simply crossing between plants grown near together, and under what would ordinarily be called uniform conditions. In order to determine the influence of crossing with fresh stock, plants of the same variety were obtained from another garden, and these were crossed with the ninth generation mentioned above. The offspring of this cross exceeded those of the other crossed plants as 100 exceeds 78 in height, as 100 exceeds 57 in the number of seed-pods, and as 100 exceeds 51 in the weight of the seed-pods. In other words, crosses from fresh stock of the same variety were nearly 30 per cent. more vigorous than crosses between plants grown side by side for some time, and over 44 per cent. more vigorous than plants from self-fertilized seeds. It is evident, from all these figures, that nature desires crosses between plants, and, if possible, between plants grown under somewhat different conditions. All the results are exceedingly interesting and important, and there is every reason to believe that, as a rule, similar results can be obtained with all plants.

Darwin extended his investigations to many plants, only a few of which need be discussed here. Cabbage gave pronounced results. Crossed plants were to self-fertilized plants in weight as 100 to 37. A cross was now made between these crossed plants and a plant of the same variety from another garden, and the difference in weight of the resulting offspring was the difference between 100 and 22, showing a gain of over 350 per cent. due to a cross with fresh stock. Crossed lettuce-plants exceeded uncrossed in height as 100 exceeds 82. Buckwheat gave an increase in weight of seeds as 100 to 82, and in height of plant as 100 to 69. Beets gave an increase in height represented by 100 and 87. Maize, when full-grown, from crossed and uncrossed seeds, gave the differences in height between 100 and 91.   Canary-grass gave similar results.

I have obtained results as well marked as these upon a large and what might be called commercial scale.   I raised the plants during the first genera-tion of seeds from known parentage, the flowers from which they came having been carefully pollinated by hand.   In some instances the second generations were grown from hand-crossed seeds, but in other cases the second generations were grown from seeds simply selected from the first-year patches.   As the experiments have been made in the field and upon a somewhat extensive scale, it was not possible to measure accurately the plants and the fruits from individuals in all cases ; but the results have been so marked as to admit of no doubt as to their character.   In 1889, several hand-crosses were made among egg-plants.   Three fruits matured, and the seeds from them were sown in 1890.   Some 200 plants were grown, and they were characterized throughout the season by great sturdiness and vigor of growth.   They grew more erect and taller than other plants near by grown from commercial seeds.   They were the finest plants which I had ever seen.   It was impossible to determine productiveness, from the fact that our seasons are too short for egg-plants, and only the earliest flowers, in the large varieties, perfect their fruit, and the plant blooms continuously through the season.   In order to determine how much a plant will bear, it must be grown until it ceases to bloom.   When frost came, I could see little differ-ence in productiveness between these crossed plants and commercial plants. A dozen fruits were selected from various parts of this patch, and in 1891 about 2,500 plants were grown from them.   Again the plants were remark-ably robust and healthy, with fine foliage, and they grew erect and tall—an indication of vigor.   They were also very productive, but as the cross had been made between unlike varieties and they were therefore unlike either parent, I could not make an accurate comparison.   But they compared well with commercial egg-plants, and I am satisfied that they would have shown themselves to be more productive than common stock could they have grown a month or six weeks longer.   Professor Munson, of the Maine Agricultural College, grew some of this crossed stock this year (1891) and he writes me that it is better than any commercial stock in his gardens.

In extended experiments in the crossing of pumpkins, squashes and gourds, carried on during several years, increase in productiveness due to crossing has been marked in many instances.   Marked increase in productiveness has been obtained from tomato crosses, even when no other results of crossing could be seen.

Bearing in mind these good influences of crossing, let us recall another series of facts following the simple change of seed.   Almost every farmer or gardener at the present day feels that an occasional change of seed results in better crops, and there are definite records to show that such is often the case. In fact, I am convinced that much of the rapid improvement in fruits and vegetables in recent years is due to the practice of buying plants and seeds so largely of dealers, by means of which the stock is often changed. Even a slight change, as between farms or neighboring villages, sometimes produces marked results, as more vigorous plants and often more fruitful ones. We must not suppose, however, that because a small change gives a good result, a violent or very pronounced change gives a better one. There are many facts on record to show that great changes often profoundly influence plants, and when such influence results in lessened vigor or lessened productiveness we call it an injurious one. Now, this injurious influence may result even when all the conditions in the new place are favorable to health and development of the plant; it is an influence which is wholly independent, so far as we can see, of any condition which interferes injuriously with the simple processes of growth. Seeds of a native physalis or husk-tomato were sent me from Paraguay in 1889 by Dr. Thomas Morong, then traveling in that country. I grew it both in the house and out-doors, and for two generations was unable to make it set fruit, even though the flowers were hand-pollinated; yet the plants were healthy and grew vigorously. The third generation, grown out-doors this year, set fruit freely. This is an instance of the fact that very great changes of conditions may injuriously affect the plant, and an equally good illustration of the power to overcome these conditions. Now there is great similarity between the effects of slight and violent changes of conditions and small and violent degrees of crossing, as both Darwin and Wallace have pointed out, and it is pertinent to this discussion to endeavor to discover if there is any real connection between the two.

It is well proved that crossing is good for the resulting offspring, because the differences between the parents carry over new combinations of characters or at least new powers into the crosses. It is a process of revitalization. And the more different the stocks in desirable characters within the limits of the variety, the greater is the revitalization; and frequently the good is of a more positive kind, resulting in pronounced characters which may serve as the basis for new varieties. In the cross, therefore, a new combination of characters or a new power fit it to live better than its parents in the conditions under which they lived. In the case of change of stock we find just the reverse, which, however, amounts to the same thing, that the same characters or powers fit the plant to live better in conditions new to it than plants which have long lived in those conditions. In either case, the good comes from the fitting together of new characters or powers and new environments. Plants which live during many generations in one place become accustomed to the place, thoroughly fitted into its conditions, and are in what Mr. Spencer calls a state of equilibrium. When either plant or conditions change, new adjustments must take place, and the plant may find an opportunity to expand in some direction in which it has before been held back ; for plants always possess greater power than they are able to express. "These rhythmical actions or functions [of the organism]," writes Spencer, "and the various compound rhythms resulting from their combinations, are in such adjustment as to balance the actions to which the organism is subject. There is a constant or periodic genesis of forces, which, in their kinds, amounts, and directions, suffice to antagonize the forces which the organism has constantly or periodically to bear. If, then, there exists this state of moving equilibrium among a definite set of internal actions, exposed to a definite set of external actions, what must result if any of the external actions are changed ? Of course there is no longer an equilibrium. Some force which the organism habitually generates is too great or too small to balance some incident force, and there arises a residuary force exerted by the environment on the organism or by the organism on the environment. This residuary force—this unbalanced force—of necessity expends itself in producing some change of state in the organism." The good results, therefore, are processes of adaptation, and when adaptation is perfectly complete the plant may have gained no permanent advantage over its former conditions, and new crossing or another change may be necessary ; yet there is sometimes a permanent gain, as when a plant becomes visibly modified by change to another climate. Now this adaptive change may express itself in two ways, either by some direct influence upon the stature, vigor or other general character, or indirectly upon the reproductive powers by which some new influence is carried to the offspring. If the direct influences become hereditary, as observation seems to show may sometimes occur, the two directions of modification may amount, ultimately, to the same thing.

For the purposes of this discussion it is enough to know that crossing within the variety and change of stock within ordinary bounds are beneficial, that the results in the two cases seem to flow from essentially the same causes, and that crossing and change of stock combined give much better results than either one alone. These processes are much more important than any mere groping after new varieties, as I have already said, not only because they are surer, but because they are universal and necessary means of maintaining and improving both wild and cultivated plants. Even after one succeeds in securing and fixing a new variety, he must employ these means to a greater or less extent to maintain fertility and vigor. In the case of some garden crops, in which many seeds are produced in each fruit and in which the operation of pollination is easy, actual hand-crossing from new stock now and then may be found to be profitable. But in most cases, the operation can be left to nature, if the new stock is planted among the old. Upon this point Darwin expressed himself as follows: "It is a common practice with horticulturists to obtain seeds from another place having a very different soil, so as to avoid raising plants for a long succession of generations under the same conditions; but with all the species which freely intercross by the aid of insects or the wind, it would be an incomparably better plan to obtain seeds of the required variety, which had been raised for some generations under as different conditions as possible, and sow them in alternate rows with seeds matured in the old garden. The two stocks would then intercross, with a thorough blending of their whole organization, and with no loss of purity to the variety; and this would yield far more favorable results than a mere change of seeds."

But you are waiting for a discussion of the second of the great features of crossing—the summary production of new varieties. This is the subject which is almost universally associated with crossing in the popular mind, and even among horticulturists themselves. It is the commonest notion that the desirable characters of given parents can be definitely combined in a pronounced cross or hybrid. There are two or three philosophical reasons which somewhat oppose this doctrine and which we will do well to consider at the outset. In the first place, nature is opposed to hybrids, for species have been bred away from each other in the ability to cross. If, therefore, there is no advantage for nature to hybridize, we may suppose that there would be little advantage for man to do so; and there would be no advantage for man, did he not grow the plant under conditions different from nature or desire a different set of characters. We have seen that nature's chief barriers to hybridization are total refusal of species to unite, and entire or comparative seedlessness of offspring. We can overcome the refusal to cross in many cases by bringing the plant under cultivation; for the character of the species becomes so changed by the wholly new conditions that its former antipathies may be overpowered. Yet it is doubtful if such a plant will ever acquire a complete willingness to cross. In like manner we can overcome in a measure the comparative seedlessness of hybrids, but it is very doubtful if we can ever make such hybrids completely fruitful. It would appear, therefore, upon theoretical grounds, that in plants in which fruits or seeds are the parts sought, no good can be expected, as a rule, from hybridization, and this seems to be affirmed by facts.

It is evident that species which have been differentiated or bred away from each other in a given locality will have more opposed qualities or powers than similar species which have arisen quite independently in places remote from each other. In the one case, the species have likely struggled with each other until each one has attained a degree of divergence which allows it to persist ; while in the other case there has been no struggle between the species, but similar conditions have brought about similar results. These similar species which appear independently of each other in different places are called representative species. Islands remote from each other, but similarly situated with reference to climate, very often contain representative species, and the same may be said of other regions much like each other, as eastern North America and Japan. Now it follows that if representative species are less opposed than others, they are more likely to hybridize with good results ; and this fact is well illustrated in the Kieffer and allied pears, which appear to be hybrids between representative species of Europe and Japan, and I am inclined to think that the same may be found to be true of the common or European apple and the wild crab of the Mississippi valley. Various crabs of the Soulard type, which I once thought to constitute a distinct species, appear upon further study to be hybrids. We will also recall that the hybrid grapes which have so far proved most valuable, are those obtained by Rogers between the American *Vitis Labrusca* and the European wine grape ; and that the attempts of Haskell and others to hybridize associated species of native grapes have given at best only indifferent results. To these good results from hybrids of fruit-trees and vines I shall revert presently.

Another theoretical point which is borne out by practice is the conclusion that, because of the great differences and lack of affinity between parents, pronounced hybrid offspring is unstable. This is one of the greatest difficulties in the way of the summary production of new varieties by means of hybridization. It would appear also, that because of the unlikeness of parents, hybrid offspring must be exceedingly variable, but as a matter of fact, in many instances the parents are so pronouncedly different that the hybrids represent a distinct type by themselves or else they approach very nearly to the characters of one of the parents. There are, to be sure, many instances of exceedingly variable hybrid offspring, but they are usually the offspring of variable parents. In other words, variability in offspring appears to follow rather as a result of variability in parents than as a result of mere unlikeness of character. But the instability of hybrid offspring when propagated by seed is notorious. Wallace writes that "the effect of occasional crosses often results in a great amount of variation, but it also leads to instability of character, and is therefore very little employed in the production of fixed and well-marked races."

I may remark again that, because of the unequal and unknown powers of the parents, we can never predict what characters will appear in the hybrids.

This fact was well expressed by Lindley a half century ago in the phrase, "Hybridizing is a game of chance played between man and plants."

Bearing these fundamental propositions in mind, let us pursue the subject somewhat in detail. We shall find at the outset that the characters of hybrids, as compared with the characters of simple crosses between stocks of the same variety, are ambiguous, negative, and often prejudicial. The fullest discussion of hybrids has been made by Focke, and he lays down the five following propositions concerning the character of hybrid offspring :

1. "All individuals which have come from the crossing of two pure species or races, when produced and grown under like conditions, are usually exactly like each other, or at least scarcely more different from each other than plants of the same species are." This proposition, although perhaps true in the main, appears to be too broadly and positively stated.

2. "The characters of hybrids are different from the characters of the parents. The hybrids differ most in size and vigor and in their sexual powers."

3. "Hybrids are distinguished from their parents by their powers of vegetation or growth. Hybrids between very different species are often weak, especially when young, so that it is difficult to raise them. On the other hand, cross-breeds are, as a rule, uncommonly vigorous ; they are distinguished mostly by size, rapidity of growth, early flowering, productiveness, longer life, stronger reproductive power, unusual size of some special organs, and similar characteristics."

4. "Hybrids produce a less amount of pollen and fewer seeds than their parents, and they often produce none. In cross-breeds this weakening of the reproductive powers does not occur. The flowers of sterile or nearly sterile hybrids usually remain fresh a long time."

5. "Malformations and odd forms are apt to appear in hybrids, especially in the flowers."

Some of the relations between hybridization and crossing within narrow limits, are stated as follows by Darwin : "It is an extraordinary fact that with many species, flowers fertilized with their own pollen are either absolutely or in some degree sterile ; if fertilized with pollen from another flower on the same plant, they are sometimes, though rarely, a little more fertile ; if fertilized with pollen from another individual or variety of the same species, they are fully fertile ; but if with pollen from a distinct species, they are sterile in all possible degrees, until utter sterility is reached. We have thus a long series with absolute sterility at the two ends—at one end due to the sexual elements not having been sufficiently differentiated, and at the other end to their having been differentiated in too great a degree, or in some peculiar manner."

The difficulties in the way of successful results through hybridization are, therefore, these : The difficulty of effecting the cross, infertility, instability, variability, and often weakness and monstrosity of the hybrids, and the absolute impossibility of predicting results. The advantage to be derived from a successful hybridization is the securing of a new variety which shall combine in some measure the most desirable features of both parents ; and this advantage is often of so great moment that it is worth while to make repeated efforts and to overlook numerous failures. From these theoretical considerations it is apparent that hybridization is essentially an empirical subject, and the results are such as fall under the common denomination of chance. And as it does not rest upon any legitimate function in nature, we can understand that it will always be difficult to codify laws upon it.

Among the various characters of hybrid offspring, I presume that the most prejudicial one is their instability, their tendency still to vary into new forms or to return to one or the other parent in succeeding generations. It is difficult to fix any particular form which we may secure in the first generation of hybrids. At the outset we notice that this discouraging feature is manifested entirely through the medium of reproduction ; and we thereby come upon what is perhaps the most important practical consideration in hybridization, the fact that the great majority of the best hybrids in cultivation are increased by bud-propagation, as cuttings, layers, suckers, buds or grafts. In fact, I recall very few instances in this country of good undoubted hybrids which are propagated with practical certainty by means of seeds. You will recall that the genera in which hybrids are most common are those in which bud-propagation is the rule, as begonia, pelargonium, fuchsia, gladiolus, rhododendron, roses, and the fruits. This simply means that it is difficult to fix hybrids so that they will come "true to seed," and makes apparent the fact that if we desire hybrids we must expect to propagate them by means of buds. And this, too, is a point which appears to have been overlooked by those who contend that hybridization must necessarily swamp all results of natural selection ; for as comparatively few plants propagate naturally by means of buds, whatever hybrids might have appeared would have been speedily lost, and all the more, also, because, by the terms of their reasoning, the hybrids would cross with other and dissimilar forms and therefore lose their identity as intermediates. Or, starting with the assumption that hybrids are intermediates and would therefore obliterate specific types, we must conclude that they should have some marked degree of stability ; but as all hybrids tend to break up when propagated by seeds, it must follow that bud-propagation would become more and more common, and this is associated in nature with decreased seed-production. Now seed-production Is the legitimate function of flowers, and we must concede that as seed-production decreased floriferousness must have decreased, and, that, therefore, pronounced intercrossing would have obliterated the very organs upon which it depends, or have destroyed itself I

But I may be met by the objection that there is no inherent reason why hybrids should not become stable through seed-production by inbreeding, and I might be cited to the opinion of Darwin and others that inbreeding tends to fix any variety whether it originated by crossing or other means. And it is a fact that inbreeding tends to fix varieties, within certain limits, but those limits are often overpassed in the case of very pronounced crosses, whether cross-breeds or true hybrids. And if it is true, as all observation and experiment show, that sexual or reproductive powers of crosses are weakened as the cross becomes more violent, we should expect less and less possibility of successful inbreeding, for inbreeding without disastrous results is possible only with comparatively strong reproductive powers. As a matter of fact, it is found in practice that it is exceedingly difficult to fix pronounced hybrids by means of inbreeding. It sometimes happens, too, that the hybrid individual which we wish to perpetuate may be infertile with itself, as I have often found in the case of squashes. It is often advised that we cross the hybrid individual which we wish to fix with another like individual, or with one of its parents. These results are often successful, but oftener they are not. In the first place it often happens that the hybrid individuals may be so diverse that no two of them are alike ; this has been my experience in many crosses. And again, crossing with a parent may draw the hybrid back again to the parental form. So long ago as last century Kölreuter proved this fact upon nicotiana and dianthus. A hybrid between *Nicotiana rustica* and *N. paniculata* was crossed with *N. paniculata* until it was indistinguishable from it ; and it was then crossed with *N. rustica* until it became indistinguishable from that parent. Yet there is no other way of fixing a hybrid to be propagated by seeds than by inbreeding, so far as I know. Fortunately, it occasionally happens that a hybrid is stable and therefore needs no fixing.

In this connection I may cite some of my own experience in crossing egg-plants and squashes ; for although the products were not true hybrids in the strict interpretation of the word, many of them were hybrids to all intents and purposes because made between very unlike varieties, and they will serve to illustrate the difficulties of which I speak. Offspring of egg-plant crosses were grown in 1890, and upon some of the most promising plants some flowers were self-pollinated. But these self-pollinated seeds gave just as variable offspring in 1891 as those selected almost at random from the patch ; and, what was worse, none of them reproduced the parent or ''came true to seed,'' and all further motive for inbreeding was gone. My labor, therefore, amounted to nothing more than my own edification ! My experience in crossing pumpkins and squashes has now extended through five years, and although I have obtained about one thousand types not named or described, I have not yet succeeded in fixing one ! The difficulty here is an aggravated one, however. The species are so exceedingly variable that all the mongrel individuals may be unlike, so that there can be no crossing between identical stocks, and if inbreeding is attempted, it may be found that the flowers will not inbreed ! And the refusal to inbreed is all the more strange because the sexes are separated in different flowers upon the same plant. In other words, in my experience, it is very difficult to get good seeds from squashes which are fertilized by a flower upon the same vine. The squashes may grow normally to full maturity, but be entirely hollow or contain only empty seeds. In some instances the seeds may appear to be good, but may refuse to grow under the best conditions. Finally, a small number of flowers may give good seeds. I have many times observed this refusal of squashes (*Cucurbita Pepo*) to inbreed. It was first brought to my attention through efforts to fix certain types into varieties. The figures of one season's tests will sufficiently indicate the character of the problem. In 1890, 185 squash-flowers were carefully pollinated from flowers upon the same vine. Only 22 of these produced fruit, and of these, only 7, or less than one-third, bore good seeds, and in some of these the seeds were few. Now these 22 fruits represented as many different varieties, so that the ability to set fruit with pollen from the same vine is not a peculiarity of a particular variety. The records of the seeds of the 7 fruits in 1891 are as follows:

Fruit No. 1.—4 vines were obtained with 4 different types of fruit, 2 of them being white, 1 yellow and 1 black.

Fruit No. 2.—23 vines : 15 types, very unlike, 12 being white and 3 yellow.

Fruit No. 3.—2 vines : 1 type of fruit which was almost like one of the original parents.

Fruit No. 4.—32 vines : 6 types, differing chiefly in size and shape.

Fruit No. 5.—20 vines : 19 types, of which 10 were white, 8 orange, 1 striped, and all very unlike.

Fruit No. 6.—13 vines : 11 types, 8 yellow, 2 black, 1 white.

Fruit No. 7.—1 vine.

These offspring were just as variable as those from flowers not inbred, and no more likely, apparently, to reproduce the parent. These tests leave me without any method of fixing a pronounced cross of squashes, and lead me to think that the legitimate process of origination of new kinds here, as indeed, if not in general, is a more gradual process of selection, coupled, perhaps, with minor crossing.

I will relate a definite attempt towards the fixation of a squash which I had obtained from crossing. The history of it runs back to 1887, when a cross was effected between a summer Yellow Crookneck and a White Bush Scallop squash. In 1889 there appeared a squash of great excellence, combining the merits of summer and winter squashes with very attractive form, size and color, and a good habit of plant. I showed the fruit to one of the most expert seedsmen of the country and he pronounced it one of the most promising types which he had ever seen ; and as he informed me that he had fixed squashes by breeding in-and-in, I was all the more anxious to carry out my own convictions in the same direction. It is needless to say that I was very happy over what I regarded as a great triumph, and I remember that I experienced a keen feeling of satisfaction that I had been able to overcome nature's prejudices. Of course I must have a large number of plants of my new variety that I might select the best, both for inbreeding and for crossing similar types. So I selected the very finest squash, having placed it where I could admire it for some days, and saved every seed of it. These seeds were planted upon the most conspicuous knoll in my garden in 1890. It was soon evident that something was wrong ; I seemed to have everything except my squash. One plant, however, bore fruits almost like the parent, and upon this I began my attempts towards inbreeding. But flower after flower failed and I soon saw that the plant was infertile with itself. Careful search revealed two or three other plants very like this one, and I then proceeded to make crosses upon it. I was equally confident that this method would succeed. When I harvested my squashes in the fall and took account of stock, I found that the seeds of my one squash had given just as many different types as there were plants, and I actually counted 110 kinds distinct enough to be named and recognized ! Still confident, in 1891 I planted the seeds of my crosses, and as the summer days grew long and the crickets chirped in the meadows, I watched the expanding blossoms and wondered what they would bring forth. But they brought only disappointment ! My squash had taken an unscientific leave of absence and I do not know its whereabouts. And when the frost came and killed every ambitious blossom, my hopes went out and have not yet returned !

Let us now recall how many undoubted hybrids there are, named and known, among our fruits and vegetables. In grapes there are the most. There are Rogers' hybrids, like Agawam, Lindley, Wilder, Salem and Barry ; and there is some reason for supposing that Delaware, Catawba, and other varieties are of hybrid origin. And many hybrids have come to notice lately through the work of Munson and others. But it must be remembered that grapes are naturally exceedingly variable and the specific limits are not well known, and that hybridization among them lacks much of that definiteness which ordinarily attaches to the subject. In pears there is the Kieffer class. In apples, peaches, plums, cherries, apricots, quinces, currants, gooseberries, blackberries, dewberries, there are no commercial hybrids. The strawberry is doubtful. Some of the raspberries, like Caroline and Shaffer, appear to be hybrids between the red and black species. Hybrids have been produced between the raspberry and blackberry by two or three persons, but they possess no promise of economic results. Among all the list of garden vegetables—plants that are propagated by seeds—I do not know of a single authentic hybrid; and the same is true of wheat—unless the Carman wheat-rye varieties become prominent, oats, the grasses, and other farm-crops. But among ornamental plants there are many ; and it is a significant fact that the most numerous, most marked, and most successful hybrids occur in the plants most carefully cultivated and protected—those, in other words, which are farthest removed from all untoward circumstances and an independent position. This is nowhere so well illustrated as in the case of cultivated orchids, in which hybridization has played no end of freaks, and in which, also, every individual plant is nursed and coddled. For such plants the struggle for existence is reduced to its lowest terms, for it must be borne in mind that even in the garden plants must fight severely for a chance to live, and even then only the very best can persist or are even allowed to try.

I am aware that this list of hybrids is much more meager than most catalogues and trade-lists would have us believe, but I am sure that it is approximately near the truth. It is, of course, equivalent to saying that most of the so-called hybrid fruits and vegetables are myths. There is everywhere a misconception of what a hybrid is and how it comes to exist; and yet, perhaps because of this indefinite knowledge, there is a widespread feeling that a hybrid is necessarily good, while the presumption is directly the opposite. The identity of a hybrid in the popular mind rests entirely upon some superficial character, and proceeds upon the assumption that it is necessarily intermediate between the parents. Hence we find one of our popular authors asserting that because the kohlrabi bears its thickened portion midway of its stem, it is evidently a hybrid between the cabbage and turnip, which respectively bear the thickened parts at the opposite extremities of the stem ! And then there are those who confound the word hybrid with *high-bred*, and who build attractive castles upon the unconscious error. And thus is confusion confounded !

But before leaving the subject of hybridization, I must speak of the old yet common notion that there is some peculiar influence exerted by each sex in the parentage of hybrids, for I shall thereby not only call your attention to what I believe to be an error, but shall also find the opportunity to illustrate still further the entanglements of hybridization. It was held by certain early observers, of whom the great Linnæus was one, that the female parent determines the constitution of the hybrid, while the male parent gives the external attributes, as form, size and color. The accumulated experience of nearly a century and a half appears to contradict this proposition, and Focke, who has recently gone over the whole ground, positively declares that it is untrue. There are instances, to be sure, in which this old idea is affirmed, but there are others in which it is contradicted. The truth appears to be this, that the parent of greater strength or virility makes the stronger impression upon the hybrids, whether it is the staminate or pistillate parent, and it appears to be equally true that it is usually impossible to determine beforehand which parent is the stronger. It is certain that strength does not lie in size, neither in the high development of any character. It appears to be more particularly associated with what we call fixity or stability of character, or a tendency towards invariability. This has been well illustrated in my own experiments with squashes, gourds and pumpkins. The common little pear-shaped gourd will impress itself more strongly upon crosses than any of the edible squashes and pumpkins with which it will effect a cross, whether it is used as a male or female parent. Even the imposing and ubiquitous great field-pumpkin which every New Englander associates with pies, is overpowered by the little gourd. Seeds from a large and sleek pumpkin which had been fertilized by gourd-pollen produced gourds and small hard-shelled globular fruits which were entirely inedible. A more interesting experiment has been made between the handsome green-striped Bergen fall squash and the little pear-gourd. Several flowers of the gourd were pollinated by the Bergen in 1889. The fruits raised from these seeds in 1890 were remarkably gourd-like. Some of these crosses were pollinated again in 1890 by the Bergen and the seeds were sown in 1891. Here, then, were crosses into which the gourd had gone once and the Bergen twice, and both the parents are to all appearances equally fixed, the difference in strength, if any, attaching rather to the Bergen. Now the crop of 1891 still carried pronounced characters of the gourd. Even in the fruits which most resembled the Bergen, the shells were almost flinty hard and the flesh, even when thick and tender, was bitter. Some of the fruits looked so much like the Bergen that I was led to think that the gourd had largely disappeared. The very hard, but thin paper-like shell which the gourd had laid over the thick yellow flesh of the Bergen, I thought might serve a useful purpose and make the squash a better keeper. And I found that it was a great protection, for the squash could stand any amount of rough handling and was even not injured by ten degrees of frost. All this was an acquisition, and as the squash was handsome and exceedingly productive, nothing more seemed to be desired. But it still remained to have a squash for dinner. The cook complained of the hard shell, but once inside, the flesh was thick and attractive and it cooked nicely. But the flavor! Dregs of quinine, gall and boneset! The gourd was still there!

We have now seen that uncertainty follows hybridization, and in closing I will say that uncertainty also attaches to the mere act of pollination. Between some species, which are closely allied and which have large and strong flowers, four-fifths of the attempts towards cross-pollination may be successful, but such a large proportion of successes is not common, and it may be infrequent even in pollinations between plants of the same species or variety. Some of the failure is due in many cases to unskillful operations, but even the most expert operators fail as often as they succeed, in promiscuous pollinating. There is good reason to believe, as Darwin has shown, that the failure may be due to some selective power of individual plants, by which they refuse pollen which in many instances is acceptable to other plants, even of the same variety or stock. The lesson to be drawn from these facts is that operations should be as many as possible and that discouragement should not come of failure. In order to illustrate the varying fortunes of the pollinator, I will transcribe some notes from my field-book:

Two hundred and thirty-four pollinations of gourds, pumpkins and squashes, mostly between varieties of one species (*Cucurbita Pepo*) and including some individual pollinations, gave 117 failures and 117 successes. These crosses were made in varying weather, from July 28 to August 30. In some periods nearly all the operations would succeed, and at other times most of them would fail. I have always regarded these experiments as among my most successful ones, and yet but half of the pollinations "took." But you must not understand that I actually secured seeds from even all these 117 fruits, for some of them turned out to be seedless, and some were destroyed by insects before they were ripe, or were lost by accidental means. A few more than half of the successful pollinations—if by success we mean the formation and growth of fruit—really secured us seeds, or about one-fourth of the whole number of efforts.

Twenty pollinations were made between tomato-flowers, and they all failed; also 7 pollinations of red-peppers, 4 of husk-tomato, 2 of *Nicotiana affinis* upon petunia and 2 of the reciprocal cross, 12 of radish, 1 of *Mirabilis Jalapa* upon *M. longiflora* and two of the reciprocal cross, 3 *Convolvulus major* upon *C. minor* and one of the reciprocal, 1 muskmelon by squash, 2 muskmelon by watermelon and one muskmelon by cucumber.

This is but one record. Now let me give you another:

Cucumber, 95 efforts: 52 success, 43 failures.
Tomato, 43 efforts: 19 successes, 24 failures.
Egg-plant, 7 efforts: 1 success, 6 failures.
Pepper, 15 efforts: 1 success, 14 failures.
Husk-tomatoes, 45 efforts: 45 failures.
Pepino, 12 efforts: 12 failures.
Petunia by *Nicotiana affinis*, 11 efforts: 11 failures.
Nicotiana affinis by petunia, 6 efforts: 6 failures.
General Grant tobacco by *Nicotiana affinis*, 11 efforts: 8 successes, 3 failures.
Nicotiana affinis by General Grant tobacco, 15 efforts: 15 failures.
General Grant tobacco by General Grant tobacco: 1 effort, 1 success.
Nicotiana affinis by *Nicotiana affinis*, 3 efforts: 2 successes, 1 failure.
Tuberous begonia, 5 efforts: 5 successes.
Total, 312 efforts: 89 successes, 223 failures.

And now, the sum of it all is this: Encourage in every way crosses within the limits of the variety and in connection with change of stock, expecting increase in vigor and productiveness; hybridize if you wish to experiment, but do it carefully, honestly, thoroughly, and do not expect too much! Extend Darwin's famous remark to read: Nature abhors both perpetual self-fertilization and hybridization.

BIBLIOGRAPHY.

The following list, while very incomplete, will enable the student to select literature bearing upon the various questions of cross-breeding and hybridization. The literature of cross-fertilization itself—the means and adaptations by which flowers are fertilized—has been omitted. Those who desire a bibliography of this subject should consult d'Arcy Thompson's excellent list in Mueller's *Fertilization of Flowers*. In the present list I have included such references as I have found upon the subject of the immediate influence of pollen upon the resulting fruit.

1724. Dudley, P. An Observation on Indian Corn. Trans. Royal Phil. Soc. vi. (2), 204-5.

1745. Cooke, Benj. Concerning the Effect which the Farina of the Blossoms of different sorts of Apple trees had on the fruit of a neighboring Tree. Trans. Royal Phil. Soc. ix. 169.

1748. Cooke, Benj. On a Mixed Breed of Apples, from the Mixture of the Farina. Trans. Royal Phil. Soc. ix. 599.

1749. Cooke, Benj. On the Effects of the Mixture of the Farina of Apple trees; and of the Mayze or Indian Corn, etc. Trans. Royal Phil. Soc. ix. 685.

1761. Koelreuter, Joseph Gottlieb. Vorläufige Nachricht von einigen das Geschlecht der Pflanzen betreffenden Versuchen und Beobachtungen. 50 pp. Leipzig. Continued in 1763, 1764, and 1766.

1793. Sprengel, Christian Konrad. Das entdeckte Geheimniss der Natur im Bau und in der Befruchtung der Blumen. 444 pp. 25 tab. Berlin.

1806. Knight, Thomas Andrew. Observations on the Means of Producing New and Early Fruits. Trans. Royal Hort. Soc. i. 30. Reprinted in Physiological and Horticultural Papers of Thomas Andrew Knight, 172.

1809. Knight, Thomas Andrew. On the Comparative Influence of Male and Female Parents on their Offspring. Trans. Royal Phil. Soc. 1809, pt. i. 392; Phys. and Hort. Papers, 343.

1814. Knight, Thomas Andrew. An Account of two New Varieties of Cherry. Trans. Royal Hort. Soc. ii. 137.

1816. Knight, Thomas Andrew. An Account of three New Peaches. Trans. Royal Hort. Soc. ii. 214.

1817. Knight, Thomas Andrew. An Account of a Peach tree produced from the Seed of the Almond tree, with some Observations on the Origin of the Peach tree. Trans. Royal Hort. Soc. iii. 1.

1818. Herbert, W. Instructions for the Treatment of the *Amaryllis longiflora*, as a hardy Aquatic, with some Observations on the Production of Hybrid Plants, and the Treatment of the Bulbs of the Genera Crinum and Amaryllis. Trans. Royal Hort. Soc. iii. 187.

1818. Knight, Thomas Andrew. Upon the Variations of the Red Currant when propagated by Seed [Crosses of white and red currants]. Trans. Royal Hort. Soc. iii. 86.

1818. Knight, Thomas Andrew. Upon the Variations of the Scarlet Strawberry (*Frugaria Virginiana*) when propagated by Seeds. Trans. Royal Hort. Soc. iii. 207.

1818. Knight, Thomas Andrew. Description of a New Seedling Plum. Trans. Royal Hort. Soc. iii. 214.

1818. Sabine, Joseph. Observations on, and Account of, the Species and Varieties of the Genus Dahlia; with Instructions for their Cultivation and Treatment [Refers to Experiments in Crossing]. Trans. Royal Hort. Soc. iii. 217.

1818. Van Mons, J. B. Substance of a Memoir on the Cultivation and Variation of Brussels Sprouts. Trans. Royal Hort. Soc. iii. 197.

1819. Anderson, David. Account of a New Melon, with a Description of the Method by which it was obtained. Trans. Royal Hort. Soc. iv. 318.

1819. Herbert, W. On the Production of Hybrid Vegetables; with the Result of many Experiments made in the Investigation of the Subject. Trans. Royal Hort. Soc. iv. 15.

1820. Sabine, Joseph. Account of a Newly Produced Hybrid Passiflora. Trans. Royal Hort. Soc. iv. 258; also vol. v. 70 (1822).

1820. Turner, John. Observations on the Accidental Intermixture of Character of Certain Fruits. Trans. Royal Hort. Soc. v. 63.

1821. Gowen, Robt. On the Production of a Hybrid Amaryllis. Trans. Royal Hort. Soc. iv. 498.

1821. Guillemin et Dumas. Observations sur l'hybridité des Plantes en Général et particulièrement sur celle de quelques Gentianes alpines. Mém. Soc. Nat. Hist. Paris, i. 79-92.

1821. Knight, Thomas Andrew. Observations on Hybrids. Trans. Royal Hort. Soc. iv. 367; Phys. and Hort. Papers, 251.

1822. Goss, John. On the Variation in the Color of Peas, occasioned by Cross Impregnation. Trans. Royal Hort. Soc. v. 234.

1823. Gowen, Robt. Description of *Amaryllis Psittacina-Johnsoni*, a New Hybrid Variety raised by William Griffin, Esq., and recently flowered in the Collection at Highclere. Trans. Royal Hort. Soc. v. 361.

1823. Gowen, Robt. On a Hybrid Amaryllis produced between *Amaryllis vittata* and *Amaryllis regina-vittata*. Trans. Royal Hort. Soc. v. 390.

1823. Knight, Thomas Andrew. An Account of some Mule Plants. Trans. Royal Hort. Soc. v. 292 ; Phys. and Hort. Papers, 275.

1823. Knight, Thomas Andrew. Some Remarks on the Supposed Influence of the Pollen, in Cross-Breeding, upon the Color of the Seed-Coats of Plants, and the Qualities of their Fruits. Trans. Royal Hort. Soc. v. 377 ; Phys. and Hort. Papers, 278.

1823. Knight, Thomas Andrew. An Account of a New Variety of Plum, Called the Downton Imperatrice. Trans. Royal Hort. Soc. v. 381.

1823. Lindley, John. A Notice of Certain Seedling Varieties of Amaryllis, presented to the Society by the Hon. and Rev. William Herbert, in 1820, which flowered in the Society's Garden in February, 1823. Trans. Royal Hort. Soc. v. 337.

1824. Knight, Thomas Andrew. Observations upon the Effects of Age upon Fruit Trees of Different Kinds ; with an Account of some New Varieties of Nectarines. Trans. Royal Hort. Soc. v. 384.

1826. Sargeret. Considérations sur la production des hybrides, des variantes et des variétés en général, et sur celle de la famille des Cucurbitacées en particular. Ann. Sci. Nat. Bot. viii. 294-314.

1826. Wiegmann, A. F. Ueber Bastarderzeugung im Pflanzenreiche.

1827. Hamelin, Baron. On the Hybrids Obtained by Baron Melazzo and others. Annales de la Soc. d'Hort. de Paris, i. No. 2. Oct. Abstr. in Gar. Mag. iii. 443.

1828. Sweet, R. The Permanency of Hybrids. Gar. Mag. iv. 182.

1830. Gowen. Hybrid Azaleas. Edward's Bot. Reg. 1830, No. viii. 1365, Vol. iv. 1407. Abstr. in Gar. Mag. vii. 62, 471.

1830. Newman, Jno. The Influence of Parent on Offspring. Gar. Mag. vi. 499.

1831. Hybrid Rhododendron. Brit. Flower Garden No. xxiii. n. s. 91. Abstr. in Gar. Mag. vii. 341.

1831. Hybrid Rhododendrons. Edward's Bot. Reg. iv. 1413. Abstr. in Gar. Mag. vii. 205.

1831. Lindley, John. Various Notes on Hybridization, in '' A Guide to the Orchard and Kitchen-Garden. London, 1831." Abstr. in Gar. Mag. vii. 579.

1831. Potentilla Russelliana. Bot. Garden No. lxxvi. 304. Abstr. in Gar. Mag. vii. 343.

1831. Remarks on Hybrids. Florists' Guide No. xliv. 174. Abstr. in Gar. Mag. vii. 205.

1831. Saunders, Wm. Hybrid Rhododendron. Gar. Mag. vii. 135.

1832. Dutrochet. The Sterility of Hybrid Plants. Gar. Mag. viii. 500.

1832. Graham, Dr. Hybrid Poppies. Gar Mag. viii. 355.

1832. Henslow, J. S. On the Examination of a Hybrid Digitalis. Gar. Mag. viii. 208. Extract from pamphlet.

1832. J. C. K. Hardihood of Hybrid Melons. Gar. Mag. viii. 52.

1832. Mallet, Robert. Hybrid Melons. Gar. Mag. vii. 87

1832. Oliver, J. A Hybrid of the Cucumber with the Maltese Melon. Gar. Mag. viii. 611.

1832. Wimmer, C. F. H. Ueber einen Bastard aus der Gattung Digitalis. Breslau Schles. Gesell. Uebersicht, 61-62 ; also 1835, 85.

1837. (Editorial). Cabbage and Horse-radish. Hovey's Mag. Hort. iii. 351.

1837. Herbert, W. Amaryllideæ, with a Treatise upon Cross-bred Vegetables. London.

1841. Wimmer, C. F. H. Ueber 6 Weidenbastarde. Breslau Schles. Gesell. Uebersicht, 93-94.

1843. Wimmer, C. F. H. Ueber die Hybridität im Pflanzenreiche. Breslau Schles. Gesell. Uebersicht, 208-209.

1844. (Editorial.) Hybridizing. Gard. Chron. 1844, 459.

1844. Gaertner, Karl Friedrich. Beiträge zur Kenntniss der Befruchtung. 644 pp. Stuttgart.

1844. Godron, D. A. Ee l'hybridité dans les végétaux. pp. 22. Nancy.

1845. L. Hybrids in Turnips. Gard. Chron. 1845, 173.

1845. Lecoq, H. Fécondation naturelle et artificiel du végétaux. Paris. Second edition, 1862.

1845. O. Hybrid between a Yellow Picotee and a Red Picotee. Gard. Chron. 1845, 363.

1847. Herbert, W. On Hybridization amongst Vegetables. Journ. London Hort. Soc. ii. 1, 81.

1847. Morton, S. G. Hybrid Plants. Am. Journ. Sci. and Arts, 2 ser. iii. 209.

1847. Wimmer, C. F. H. Ueber die Hybridität der Weiden. Breslau, Schles. Gesell. Uebersicht, 124-131.

1847-8. Regel, E. Ueber Varietäten und Bastarde im Pflanzenreiche. Mittheil. Zürich, i. Heft 2, 69-71.

1848. Loiseleur-Deslongchamp. Observations sur les Plantes dont les fleurs paraissent de refuser á l'Hybridation. Rev. Hort. 3 ser. ii. 149.

1848. Mackenzie, G. S. An Account of Some Hybrid Melons. Jour. London Hort. Soc. iii. 299.

1849. Gaertner, Karl Friedrich. Versuche und Beobachtungen über die Bastarderzeugung in Pflanzenreich. 791 pp. Stuttgart.

1849. Làhérard. Raisin précoce Malingre. Rev. Hort. 3 ser. ii. 444.

1849. Pépin. Hybrides des *Abutilon striatum* et *venosum*. Rev. Hort. 3 ser. iii. 46.

1849. Wimmer, C. F. H. Uebersicht der bisher bekannt gewordenen Bastarde von Salix. Breslau Schles. Gesell. Uebersicht, 87-93.

1850. Berkeley, M. J. Gaertner's Observations upon Muling among Plants. Journ. London Hort. Soc. v. 156 ; vi. 1 (1851).

1850. Naudin, C. Hybridation des Orchidées. Rev. Hort. 3 ser. iv. 9.

1850. Standish and Noble. A Chapter in History of Hybrid Rhododendrons. Journ. London Hort. Soc. v. 271.

1851. Decaisne, J. Hybridation. Rev. Hort. 3 ser. v. 62.

1852. Rousselon and others. Sur l'Hybridation. Ann. d'Hort. de Paris, xliii. 35.

1852. Weddell, H. A. Description d'un cas remarqueble d'hybridité, entre des Orchidées de genres différents. Ann. des Sci. Nat. Bot. 3 ser. xviii. 5.

1853. Grenier, Ch. De l'hybridité et de quelques hybrides en particulier. Ann. des. Sci. Nat. Bot. 3 ser. xix. 140.

1853. Malbranche, M. A. De l'Origine des Espèces en Botanique, et de l'Apparition des Plantes sur le Globe.

1853. Morren, C. La Fécondation des Céréales. Liège.

1853. Wimmer, C. F. H. Wild-Wachsende Bastardpflanzen, hauptsächlich im Schlesien beobachtet. Breslau Schles. Gesell. Denkschr. 143-182.

1854. Klotzsch, J. F. Ueber die Nutzanwendung der Pflanzen-Bastarde und Mischlinge. Berlin, Bericht. pp. 535-562.

1855. (Editorial.) Hybrids Exhibited at the Horticultural Society. Gar. Chron. 1855, 451.

1855. Godron, D. A. De la Fécondation des Ægylops par les Triticum. Nancy.

1855. Klotzsch. Sur l'Utilité des Hybrides. Rev. Hort. 4 ser. iv. 342.

1855. Naudin, C. Hybridation des Cucurbitacées. Rev. Hort. 4 ser. iv. 64.

1855. Naudin, C. Réflexions sur l'Hybridation dans les Végétaux. Rev. Hort. 4 ser. iv. 351.

1855. Regel, E. Zur Ægilops Frage. Bot. Zeit. xiii. 569-573.

1856. Gray, Asa. Review of Hooker and Thomson's Flora Indica, in which Hybridization is discussed in its relation to Variability. Am. Jour. Sci. and Arts. 2 ser. xxi. 135 ; reprinted in Scientific Papers of Asa Gray, i. 62.

1856. Naudin, C. Nouvelles recherches sur les charactères spécifiques et

les variétés des plantes du genre Cucurbita. Ann. des Sci. Nat. Bot. 4 ser.
v. i5. Review by Asa Gray in Am. Journ. Sci. and Arts, 2 ser. xxiv. 440;
reprinted in Scientific Papers of Asa Gray, 1. 83.

1856. Radlkofer, L. Befruchtung der Phanerogamen. Leipzig.

1856. Regel, E. Der Bastard zwischen *Ægilops ovata* and *Triticum vulgare*. Bonplandia, iv. 243–246.

1856. Sansey, J. De l'Hybridation. Rev. Hort. 4 Ser. v. 223.

1857. Regel, E. Ueber Parthenogenesis and Pflanzen Bastarde. Bonplandia, v. 302–305.

1857. Vilmorin, L. and J. Groenland. Note sur l'Hybridation du Genre Ægilops. Rev. Hort. 193.

1858. Gray, Asa. Action of Foreign Pollen upon the Fruit. Am. Journ. Sci. and Arts, 2 ser. xxv. 122.

1858. Regel, E. Hybridation des *Begonia rubrovenia* et *xanthina*. Gartenflora, Jan. 1858. Ext. in Ann. d'Hort. de Paris, iv. 442.

1859. (Editorial.) A Leguminous Hybrid. Gard. Chron. 1859, 71.

1859. Darwin, Charles. Chapter on Hybridism (ix.) in "Origin of Species."

1859. Saulget, G. du. Résultats de l'Hybridation du *Cactus* (*Epiphyllum*) *Ackermanni*. Ann. d'Hort. de Paris v. 394.

1859. Stange, F. E. Hybrides et formes diverses obtenus par la Fécondation de differents Begonia entre eux. Hamburger Garten-und Blumenzeitung. March, 1859. Reviewed in Ann. d'Hort. de Paris v. 295.

1859. Vilmorin, L. Notices sur l'Amélioration des plantes par le semis et. Considerations sur l'Hérédité dans les Végétaux. Paris. Review by Asa Gray in Am. Journ. Sci. and Arts, 2 ser. xxvii. 440; reprinted in Scientific Papers of Asa Gray, i. 109.

1860. Wylie, Dr. Hybridizing the Peach with the Plum. Gard. Month. ii. 231.

1861. Anderson-Henry. I. Variegation, Cross-Breeding and Muling of Plants. Journ. Hort. xxvii. o. s. 41.

1861. Beaton, D. Cross-Breeding and Hybridizing. Journ. Hort. xxvii. 237.

1861. Beaton, D. Facts and Opinions, Past and Present, relative to Cross-Breeding Plants. Journ. Hort. xxvii. o. s. 289.

1861. Beaton, D. Phenomena in the Cross-Breeding of Plants. Journ. Hort. xxvi. o. s. 112.

1861. Brent, B. P. Cross-Breeding Sweet-Peas. Journ. Hort. xxvi. o. s. 160.

1861. Carr. Quelques mots sur les Hybrides. Rev. Hort. 47.

1861. Darwin, Chas. Cross-Breeding in Plants. Journ. Hort. xxvi. o. s. 151.

1861.   Naudin, C.   Sur les plantes Hybrides.   Rev. Hort. 396.

1861.   Smith, Wm.   Hybridizing Mangle's Silver-Bedding Geranium.   Journ. Hort. xxvii. o. s. 195.

1861.   Wimmer, C. F. H.   Ueber Weiden-Bastarde und über *Salis grandi-flora.*   Breslau Schles. Gesell.   Jahresb. xxxix. 100-101.

1862.   Beaton, D.   Progress of Cross-Breeding among Florists' Flowers. Journ. Hort. xxvii. o. s. 309.

1862.   Caspary, Robt.   Ein Bastard von *Digitalis purpurea*, L., und *lutea*, L. Schrift. Phys. Ökon. Gesell. Königsb. iii. 139-146.

1862.   Goeze, Emmanuel.   Hybridation du *Poa aquatica* et du *Clyceria fluitans.*   Rev. Hort. 375.

1862.   Lecoq.   See 1845.

1863.   Anderson, J.   On Orchid Cultivation, Cross-Breeding and Hybridizing.   Journ. Hort. 206.

1863.   Brongniart, Decaisne, Tulasne, Moquin-Tandon, and Ducharte.   Rapport sur la question de l'Hybridité dans les Végétaux, mise an Concours par l'Académie des Sciences en 1861.   Ann. des Sci. Nat. Bot. 4 ser. xix. 125.   See also Ann. d'Hort. de Paris, ix. 56.

1863.   Godron, D. A.   Des hybrides Végétaux considérés au point de vue de leur fécondité et de la perpétuité ou non-perpétuité de leurs caractéres. Ann. des Sci. Nat. Bot. 4 ser. xix. 135.

1863.   Godron, D. A.   Recherches expérimentales sur l'hybridité dans le règne végétal, pp. 76.   Nancy.

1863.   Naudin, C.   Nouvelles recherches sur l'hybridité dans les végétaux. Ann. des Sci. Nat. Bot. 4 ser. xix. 180.

1864-66.   Marès, Henri.   De la fécondation des.fleurs steriles de la Vigne. Mém. Acad. Sci. Montpellier, vi. pp. 40-42.

1865.   Bentham, George.   Anniversary address to the Linneæan Society. Journ. Linn. Soc. viii. pp. xi-xxiii.

1865.   Carrière, E. A.   Production et Fixation des variétiés dans les végétaux.   Pp. 72.   Paris.

1865.   Déy.   Lettres sur la fécondation artificielle des plantes.   Vesoul.

1865.   (Editorial.)   Is the Quality of Fruit Changed by Hybridizing?   Gar. Month. vii. 304 ; vol. viii. 144.

1865.   Gray, Asa.   Spontaneous Return of Hybrid Plants to their Parental Forms [Note upon the essays of Naudin and Godron].   Am. Journ. Sci. and Arts, 2 ser. xxxix. 107.

1865.   McNabb.   Will Pines Hybridize ?   Gar. Month. vii. 318.

1865.   Mendel, G.   Versuche über Pflanzen-Hybriden.   Brünn Verhandl. iv. 3-47.

1865.   M. J. B.   Variation in Hybrids.   Gar. Chron. 1865, 601.

1865. Nägeli, Karl. Die Bastardbildung im Pflanzenreiche. Sitzungsber. der Köngl-bayer. Akad, der Wissensch. zu München, ii. 395-443.

1865. Scott, John. Notes on the Sterility and Hybridization of Certain Species of Passiflora, Disemma, and Tacsonia. ,Journ. Linn. Soc. Bot. viii. pp. 197-206.

1865. Verlot, J. B. Sur la production des Variétiés.

1865. Wichura, Max. Die Bastard-befruchtung im Pflanzenreich erläutert an den Bastarden der Weiden. 95 pp. 2 tab. Breslau.

1866. Godron, D. A. Nouvelles recherches sur l'Hybridité dans le règne végétal. pp. 40. Nancy.

1866. Neumann, L. Phénomène d'hybridation observé dans le genre Mathiola. Rev. Hort. 286.

1866. Standish, J. Hybridizing Fruits. Gar. Chron. 1866, 109. Also Gar. Month. ix. 284.

1866. Stayman, J. Is the Quality of Fruit Changed by Hybridization? Gar. Month. viii. 101.

1867. Anderson-Henry, I. The Hybridization of Plants. Gar. Month. 1867, 379.

1867. Anderson-Henry, I. On Pure Hybridization, or Crossing Distinct Species of Plants. Gar. Chron. 1867, 1296, 1313.

1867. Braun, A. Rejuvenescence, Especially in Mosses and Ferns. Gar. Month. ix. 147.

1867. Crucknell, Chas. Does Hybridizing Change the Fruit as well as its Progeny? Gar. Month. ix. 165.

1867. (Editorial.) Hybridization. Gar. Chron. 1867, 403.

1867. Hildebrand, F. Die Geschlechtsvertheilung bei den Pflanzen und das Gesetz der Vermiedenen und unvortheilhaften stetigen Selbstbefruchtung. Pp, 92. 62 fig.

1867. Marès, H., and J. Planchon. Sur la floraison et la fructification de la vigne. Comp. Rend. lxiv. pp. 254-259. Translation in Ann. and Mag. Nat. Hist. 3 ser. xix. pp. 220-224.

1867. Nägeli, K. Intermediate Forms in Plants. Gar. Chron. 1867, 405.

1867. R. D. Hybrid Tropæolums. Gar. Chron. 1867, 906.

1867. Rivers, Thos. Hybridizing and Cross-Breeding. Gard. Chron. 1867, 516.

1867. Stelzner. The Hybridization of Ferns. Gard. Month. ix. 60.

1868. Darwin, Charles. Chapters xv.-xix. on "Variation of Animals and Plants under Domestication."

1868. (Editorial.) Hybrid Coleuses. Gar. Chron. 1868, 1210.

1868. Hildebrand, F. Ueber die Einfluss des fremden Pollens auf die Beschaffenheit der durch ihn erzeugten Frucht. Leipzig.

1868.   Moore, Jacob.   Grape Hybrids.   Gard. Month. x. 240.

1868.   St. Pierre, M. G. de.   Hybridization of the Gourd.   Gar. Chron. 1868, 681.

1868.   Tidey, A.   Effect of Cross-Impregnation on Seeds.   Journ. Hort. xxxix. o. s. 199.

1868.   T. M.   New Hybrids of Coleus.   Gar. Chron. 1868, 376.

1868.   Wylie, A. P.   Hybrid Grapes.   Gar. Month. x. 153.

1869.   Darwin, Chas.   On the Character and Hybrid-like Nature of the Offspring from the Illegitimate Unions of Dimorphic and Trimorphic Plants.   Jour. Linn. Soc. Bot. x. pp. 393-437.

1869.   Darwin, Charles.   On the Specific Differences between *Primula veris, P. vulgaris* and *P. elatior;* and on the Hybrid Nature of the Common Oxlip.   With Supplementary Remarks on Naturally-produced Hybrids in the genus verbascum.   Jour. Linn. Soc. Bot. x. pp. 437-454.

1869.   Mendel, G.   Ueber einige aus Künstlicher Befruchtung Gewonnenen Hieracium-Bastarde.   Brünn Verhandl. viii. 26-31.

1869.   Wilder, Marshall P.   Remarks on Hybridization.   Gar. Month. xi. 3.

1870.   Anderson-Henry, I.   Imperfect Hybridity.   Gar. Chron. 1870, 1634; Journ. Hort. xliv. 493.

1870.   Higbee, C. H.   Hybrid Grapes.   Gar. Month. xii. 205.

1870.   Meehan, Thos.   Cross-Fertilization and the Law of Sex in Euphorbia.   Gar. Month. xii. 262.

1871.   Anderson-Henry, I.   Hybridism vs. Mimicry.   Gar. Chron. 1871, 10.

1871.   Campbell, G. W.   Grape Seedlings and Hybrids.   Gar. Month. xiii. 80.

1871.   Kerner, A.   Koennen aus Bastarden Arten werden?   Vienna.

1871.   Saporta et Marion.   Observations sur un Hybride spontané du Terebinthe et du Lentisque.   Paris.

1871.   Wylie, A. P.   Curious Results of Hybridization on Pollen.   Gard. Month. xiii. 123.

1872.   Anderson-Henry, I.   Hybridism vs. Mimicry.   Gar. Chron. 1872, 671.

1872.   Anderson-Henry, I.   On Pure Hybridization, or Crossing Distinct Species of Plants.   Garden i. 480, 506, 521, 574.

1872.   Blomberg, A,   Om Hybridbildning hos Phanerogamae.   Stockholm.

1872.   Carrière, E. A.   Des Hybrides.   Rev. Hort. 51.

1872.   Denny, J.   The Relative Influence of Parentage in Cross-Breeding.   Garden ii. 15.

1872.   Denny, J.   The Influence of Parent on Offspring.   Gar. Chron. 1872, 1263.

1872. (Editorial.) The Influence of Parentage on the Offspring. Gar. Chron. 1872, 1321.

1872. (Editorial.) Hereditary Transmission. Gar. Chron. 1872, 1224.

1872. Grieve, P. Hybridization. The Relative Influence of Parentage in Flowering Plants (Pelargonium). Gar. Chron. 1872, 1103.

1872. Laxton, Thomas. Notes on Some Changes and Variations in the Offspring of Cross-Fertilized Peas. Jour. Roy. Hort. Soc. new ser. iii. pp. 10-13.

1872. Maximowicz, C. J. Einfluss fremden Pollens auf die Form der erzeugten Frucht. Bull. de l'Acad. Imper. des Sci. St. Petersburg, viii. 422.

1872. Moore, Jacob. Immediate Effect on Hybridized Corn. Gar. Month. xiv. 218.

1872. Pearson, J. W. The Relative Influence of Parentage in Cross-Breeding. Garden ii. 51.

1872. Wylie, A. P. Experiments in Hybridization, Especially with Peaches and Nectarines. Gar. Month. xiv. 302.

1873. Anderson-Henry, I. Imperfect Hybridity. Jour. Hort. xlix. o. s. 185.

1873. Arnold, C. Immediate Effects of Cross-Fertilization. Gar. Month. xv. 104.

1873. (Editorial.) New Hybrid Rose. Floral Mag. 1873, No. 14, n. s.

1873. Jordan, A. Remarques sur le fait de l'existence en société, a l'état sauvage des espèces végétales affines, et sur d'Autres faits relatifs a la question de l'espèce. French Assoc. Adv. Sci. Lyons, 1873.

1873. Meehan, T. Influence of Pollen in Cross-Fertilization. Phila. Acad. Sci. 1873, pp. 16, 17.

1873. Saunders, W. Hybrid Raspberries. Gar. Month. xv. 314.

1873. T. Hybridization on Indian Corn. Garden iv. 10.

1873. Temlin, L. J. Hybridization and Cross-Fertilization. Gar. Month. xv. 302.

1873. Wilder, Marshall P. Hybridization. Gar. Chron. 1873, 575.

1874. Arnold. Influence of Parent on Offspring. Garden v. 222.

1874. (Editorial.) Hybridation et fécondation pratiquées par M. Rivers. Rev. Hort. 182.

1874. (Editorial.) New Blandfordias. Floral Mag. 1874, No. 32 n. s.

1874. (Editorial.) New Variety of Amaryllis. Flor. Mag. 1874, No. 31, n. s.

1874. Experiments in Hybridizing. Journ. Hort. li. o. s. 442.

1874. Gray, Asa. Do Varieties Wear Out? New York Tribune, Semi-Weekly ed. Dec. 8, 1874; reprinted in Am. Journ. Sci. and Arts, 3 ser. ix. 109, and in Scientific Papers of Asa Gray, ii. 174, and Darwiniana, chapter xii.

1875. Barron, A. F. Hybridization of the Monukka and Black Hamburgh Grapes. Garden vii. 19.

1875. Cox, E. W. Heredity and Hybridism. A Suggestion. London.

1875. Devansaye, A. de la. Fécondation et Hybridation des Aroidées. Bull. Soc. Hort. Maine-et-Loire, 223; Fl. des Serres, xxii. 37-47.

1875. (Editorial.) A New Hybrid Lily. Floral Mag. 1875, No. 48 n. s.

1875. (Editorial.) A Hybrid Tacsonia. Gar. Chron. iv. n. s. 167.

1875. (Editorial.) Hybrid Fern. Gar. Month. xvii. 330.

1875. (Editorial.) Variation in Hybrid Plants. Gar. Chron. iv. n. s. 748.

1875. Grieve, P. Results of Hybridization, Especially in Pelargoniums. Gar. Chron. iv. n. s. 711.

1875. Koch, Karl. Hybrid Aroids. Gar. Chron. iv. n. s. 398.

1875. Lowe, E. J. Hybrid Pelargoniums. Gar. Chron. iii. n. s. 83, 115.

1875. Meehan, T. Are Insects any Material Aid to Plants in Fertilization ? Proc. A. A. A. S. xxiv. 243-251; Gar. Chron. iv. 327; Archiv. des Sci. Nat. lvi. pp. 294-296 (1876).

1875. Naudin, C. Variation disordonnée des plantes hybrides et déductions qu-on peut en tirer. Ann. des Sci. Nat. Bot. 6 ser. ii. 73; Comp- Rend. 1875 (2), 520, 553. Review by Asa Gray in Am. Jour. Sci. and Arts, 3 ser. xi. 153; reprinted in Scientific Papers of Asa Gray, i. 212.

1875. Parkman. A New Hybrid Lily. Gar. Chron. iv. n. s. 237, 366.

1875. Sisley, J. Hybridization, Especially in Pelargoniums. Gar. Chron. iv. n. s. 654.

1875. Tillery, Wm. Hybridizing Liliums. Gar. Chron. iv. n. s. 401.

1875. Tillery, Wm. Results of Crossing the Black Monukka Grape. Gar. Chron. iii. n. s. 84.

1875. Wilson, A. S. Wheat and Rye Hybrids. Gar. Chron. iii. n. s. 496.

1876. A. D. The Hybridization of the Potato. Garden x. 419.

1876. Anderson-Henry, I. A Hybrid between *Fuchsia virgata* and *F. procumbens.* Gar. Chron. vi. n. s. 592.

1876. B. Hybridization in Relation to Grafting. Garden x. 349.

1876. Bessey, C. E. Immediate Effects of Cross-Fertilization on Fruits. Gardener's Monthly, 23.

1876. Burbidge, F. W. Hybrids, Especially a Hybrid between the Red Cedar and the American Arbor-vitæ. Gar. Chron. v. n. s. 373.

1876. Crossed Auriculas. Floral Mag. No. 54, n. s.

1876. Darwin, Charles. The Effects of Cross and Self-Fertilization in the Vegetable Kingdom. London. Review by Asa Gray in Am. Jour. Sci. and Arts, 3 ser. xiii. 125; reprinted in Scientific Papers of Asa Gray, i. 217. Review by W. J. Beal, Rep. Mich. Pom. Soc. 1877, 454.

1876. Devansaye, A. de la. Fructification des Aroidées. Rev. Hort. 288-9.

1876. (Editorial.) A New Clematis. Floral Mag. 1876, No. 59, n. s.

1876. (Editorial.) Hybrid Columbines. Floral Mag. 1876, No. 60, n. s.

1876. (Editorial.) Hybrid Narcissi. Floral Mag. 1876, No. 56, n. s.

1876. (Editorial.) Immediate Effects of Cross-Fertilization on Fruits. Gar. Month. xviii. 23.

1876. (Editorial.) Peach and Apricot Hybrids. Gar. Month. xviii. 214.

1876. Grieve, P. Influence of Foreign Pollen on the Progeny of Plants. Gard. Chron. v. 699 ; vi. 49.

1876. Grieve, P. The Sterility of Hybrids. Garden x. 67, 150.

1876. Hovey, C. M. The Sterility of Hybrids. Garden x. 251.

1876. Meehan, T. Are Insects a Material Aid in Fertilization? Garden x. pp. 493-494.

1876. Pasquale, G. A. Fecondita d'un Mandarino. Naples.

1876. Syme, Geo. Hybridization. Gar. Chron. v. n. s. 404.

1877. Beal, W. J. Experiments in Breeding. Rep. Mich. Bd. Agric. 1877, 55-56.

1877. Burbidge, F. W. Pp. 87-168 of "Cultivated Plants, their Propagation and Improvement." London.

1877. Creighton, J. H. Seedlings of Hybrid Grapes. Gar. Month. xix. 113.

1877. (Editorial.) Hybrid Greenhouse Rhododendrons. Floral Mag. 1877, No. 63, n. s.

1877. (Editorial.) Various Hybrids. Floral Mag. 1877, No. 67, 71, n. s. ; 1879, No. 89. n. s.

1877. Haskell, Geo. An Account of Various Experiments for the Production of New and Desirable Grapes. Pp. 18. Ipwich, Mass.

1877. Henslow, Geo. The Fertilization of Plants. Gar. Chron. pp. 42, 139, 203, 270, 336, 534, 560. (Review of Darwin's Cross- and Self-Fertilization.)

1877. Meehan, Thos. Mr. Darwin on the Fertilization of Flowers. Penn. Monthly, June.

1878. Barron, A. F. Hybridization of the Monukka and Black Hamburgh Grapes. Gard.Month. xx. 16.

1878. Beal, W. J. Cross-Breeding of Fruits. Rep. Mich. Pom. Soc. 1878, 51.

1878. Beal, W. J. The Improvement of Grains, Fruits and Vegetables. Rep. Mich. Bd. Agric. 1878, 445.

1878. (Editorial.) Parentage in Hybrids. Gar. Chron. x. n. s. 50.

1878. Heckel, Ed. Des relations que présentent les phénomènes propres aux organes reproducteurs de quelques phanérogames avec la fécondation croisée et la fécondation directe. Comp. Rend. lxxxvii. pp. 697-700.

1878. Meehan, Thos. On Hybrids in Nature (Extract). Gar. Month. xx. 344.

1878. Parkman, F. Hybridization of Lilies. Bull. Bussey Inst. ii. 161.

1878-9. Eaton, D. C. Hybrids and Hybridism, in Rep. Conn. Board Agr. 34.

1879. Beal, W. J. Experiments in Cross-Breeding Plants of the Same Variety. Am. Journ. Arts and Sci. ser. iii. xvii. pp. 343-345.

1879. Beal, W. J. The Effect of Pollen on Corn during the First Year. Rep. Mich. Bd. Agric. 1879, 198.

1879. (Editorial.) Deux hybrides intéressantes. Rev. Hort. 283.

1879. Henslow, Geo. On the Self-fertilization of Plants. Trans. Linn. Soc. Bot. 2 ser. i. Review by Asa Gray in Am. Journ. Sci. and Arts, 2 ser. xvii. 489; reprinted in Scientific Papers of Asa Gray, i. 263.

1879. Wallace, Dr. On Hybridization of Lilies, in "Notes on Lilies and their Culture," 60-65 (2 ed. 1879.) London.

1880. Beal, W. J. Crossing Corn with Foreign Stock. Rep. Mich. Bd. Agric. 1880, 287.

1880. (Editorial.) A Hybrid Fir. Gar. Month. xxii. 39.

1880. (Editorial.) Results of Hybrization, especially with Regard to Pelargoniums. Gar. Chron. xiv. n. s. 391.

1880. Ernst, Dr. The Relative Degrees of Fertility of *Melochia parviflora* when Intercrossed. Gar. Chron. xiii. n. s. 48.

1880. Haberlandt, G. Die Samen-production des Rothklees. Biedermann's Centralbl. für Agriculturchemie, pp. 199-201.

1880. Henslow, Geo. Self-fertilization as a Cause of Doubling. Gar. Chron. 534.

1880. Laxton, T. T. Cross-fertilizing and Raising Roses from Seed in England. Gar. Month. xxii. 322.

1880. Martindale, I. C. Sexual Variation in *Castanea Americana*. Phil. Acad. Sci. 1880, pp. 351-353.

1880. Meehan, T. On the Laws Governing the Production of Seed in *Wistaria Sinensis*. Journ. Linn. Soc. Bot. xvii. pp. 90-92.

1880. Some Phenomena Attached to Hybridization. Floral World, 1880, 220.

1881. Briggs, T. R. A. On the Production of Hybrids in the Genus Epilobium. Journ. of Botany, xix. 308.

1881. Focke, W. O. Die Pflanzenmischlinge. Pp. 569. Berlin.

1882. (Editorial.) Hybrid Orchids. Gar. Month. xxiv. 184.

1882. (Editorial.) Hybrid Nepenthes. Gar. Month. xxiv. 371.

1882. Lazenby, W. R. Improvements or Modifications of Varieties by Crossing or Hybridizing. Rep. O. Exp. Sta. i. 66-68.

1883. B. J. C. Effects of Cross-Fertilization on Fruit. Gar. Month. xxv. 274.

1883. Carman, E. S. Cross-Breeding Wheat. Gar. Month. xxv. 52.

1883. Crucknell, Chas. Hybridizing Araceous Plants. Gar. Month. xxv. 182.

1883. Culverwell, W. A Hybrid between the Gooseberry and the Black Currant. Gar. Chron. xix. n. s. 635.

1883. (Editorial.) A Hybrid between the Strawberry and the Raspberry. Gar. Chron. xx. n. s. 12.

1883. (Editorial.) Hybrids. Gar. Month. xxv. 184.

1883. (Editorial.) The Hybrid Cotton Plant. Gar. Month. xxv. 23.

1883. Lazenby, W. R. Crossing Varieties of Corn. Rep. O. Exp. Sta. ii. 63-65.

1883. Sturtevant, E. L. Corn Hybridization. Rep. N. Y. Exp. Sta. i. (for 1882), 54.

1884. B. T. The Fertility of Hybrids. Gar. Chron. xxii. n. s. 406.

1884. Carman, E. S. Hybrids between Wheat and Rye. Gar. Month. xxvi. 278.

1884. (Editorial.) A New Hybrid Pear. Gar. Month. xxvi. 142.

1884. (Editorial.) Hybrid Anthuriums. Gar. Month. xxvi. 330.

1884. (Editorial.) Hybrid Lobelias—Illustrating Fertility of Hybrids. Gar. Month. xxvi. 356.

1884. (Editorial.) Hybrid Orchids—Illustrating Fertility of Hybrids. Gar. Month. xxvi. 279.

1884. (Editorial.) Hybrid Sarracenias. Gar. Month. xxvi. 53.

1884. (Editorial.) Immediate Effect of Crossing on Fruit. Gar. Month. xxvi. 305.

1884. Garfield, C. W. Influence of Pollen. Mich. Hort. Soc. 246.

1884. Lazenby, W. R. Corn Experiments. Rep. O. Exp. Sta. iii. 64.

1884. Meehan, Thomas. Immediate Influence of Pollen on Fruit. Proc. Phila. Acad. Nat. Sci. 297.

1884. Meehan, Thos. The Fertility of Hybrids. Gar. Chron. xxii. n. s., 362, 394.

1884. Pearson, C. E. Results obtained from Hybridizing Pelargoniums. Gar. Chron. xxi. n. s. 712.

1885. Arnold, Charles. Hybrid Raspberries. Amer. Gar. vi. 50.

1885. Fertile Hybrids. Gar. Month. xxvii. 250.

1885. Goff, E. S. Results of Cross-fertilization in Peas. Rep. N. Y. Exp. Sta. iii. (for 1884) 236.

1885. Lazenby, W. R. Experiments with Corn : Cross-fertilization. Rep. O. Exp. Sta. iv. 31.

1885. Lazenby, W. R., T. J. Burrill, A. S. Fuller, E. Williams, *et al.*
. Upon the Immediate Influence of Pollen, especially upon the Strawberry.
Proc. Amer. Pom. Soc. 66 *et seq.*

1885. Meehan, Thomas. On the Immediate Action of Pollen on Fruit.
Rural N.-Y. June 13 and 20. See also Gar. Month. 1885, pp. 86, 116, 150,
341.

1885. Munson, T. V. American Grapes [with discussion of Hybridization].
Proc. Amer. Pom. Soc. 95.

1885. Sturtevant, E. Lewis. An Observation in the Hybridization and
Cross-Breeding of Plants. Amer. Nat. xix. 1041.

1885. Sturtevant, E. L. Hybrid Barley. Rep. N. Y. Exp. Sta. iii. (for
1884), 81.

1885. Sturtevant, E. L. Crossing in Maize. 3rd Rep. N. Y. Exp. Sta.
(for 1884), 148.

1885. Veitch, H. J. The Hybridization of Orchids. Gar. Chron. xxiii. n.
s. 628 ; Garden, xxvii. 450.

1886. Carman, E. S. Blackberry and Raspberry Hybrids. Rural N.-Y.
53 (Jan. 23) ; also 468 (July 17).

1886. Carman, E. S. Wheat-Rye Hybrids. Rural N.-Y. Feb. 13.

1886. (Editorial.) A Hybrid Palm. Gar. Month. xxviii. 118.

1886. (Editorial.) A New Hybrid Pear. Gar. Month. xxviii. 365.

1886. (Editorial.) Hybrid Dendrobiums. Garden xxix. 291.

1886. (Editorial.) Hybrid Montbretias. Garden xxx. 139.

1886. (Editorial.) Hybridizing the Potato. [An account of Experiments
made by Arthur W. Sutton, England.] Country Gentleman, 974 (Dec. 23).

1886. (Editorial.) Immediate Effects of Crossing on Fruit. Gar. Month.
xxviii. 148.

1886. (Editorial.) The Effects of Hybridizing in Wheat. Gar. Chron.
xxvi. n. s. 240.

1886. Goff, E. S. The Influence of Different Pollens upon the Character
of the Strawberry. 4th Rep. N. Y. Exp. Sta. (for 1885), 227.

1886. Goff, E. S. Cross-Fertilizations. Same, p. 184.

1886. Halsted, B. D. Cross-Fertilization of Fruits [Immediate Influence
of Pollen]. Bull. Bot. Dept. Iowa Agric. Coll. 37.

1886. H. G. *Lælia Batemaniana*, a Hybrid between a Cattleya and a
Sophronitis. Gar. Chron. xxvi. n. s. 263.

1886. L. C. Bigeneric Hybrids. Journ. Hort. xiv. 3 ser. 127.

1886. Munson, T. V. Hybridity and its Effects. Proc. Soc. Prom. Agric.
Sci. vii. 51.

1886. Munson, T. V. Hybridization. 3 parts. Amer. Gar. vii. 42, 75, 168.

1886. Munson, T. V. Peach and Plum. Country Gentleman, 972 (Dec. 23).

1886. Sturtevant, E. L.   Atavism the Result of Cross-Breeding in Lettuce. Proc. Soc. Prom. Agric. Sci. vii. 73.

1886. Sturtevant, E. L.   Hibridization and Cross-Breeding of Plants. [Several Kitchen-Garden Vegetables.] Am. Gar. vii. 307.

1887. Bailey, L. H.   Notes on Crossing and Hybridizing.   Bull. 31, Mich. Agric. Coll. 90.

1887. Bauer, M.   A Hybrid Cypripedium.   Gar. Chron. i. 3 ser. 419.

1887. Burrill, T. J.   Notes on Crossing in Annual Hort. Rep. in Rep. Bd. Trustees Univ. Ill. xiv.

1887. Claypole, E. W.   Secondary Results of Pollination.   Rep. U. S. Dept. Agric. 318.

1887. Crozier, A. A.   Immediate Influence of Cross-Fertilization upon the Fruit.   Rep. U. S. Dept. Agric. 312.

1887. Crozier, A. A.   Immediate Influence of Pollen.   Agric. Sci. i. 35, 227.

1887. Crozier, A. A.   Influence of Cross-Fertilization on the Fruit.   Proc. Amer. Pom. Soc. 21.

1887. Crozier, A. A.   Some Crosses in Corn.   Proc. Soc. Prom. Agric. Sci. viii. 91.

1887. Dod, Wolley.   A Hybrid between *Narcissus Bulbocodium* var. *nivalis* and *N. triandrus*.   Gar. Chron. 3 ser. i. 358.

1887. Goff, E. S.   Attempts to Obtain Potato Hybrids.   Rep. N. Y. Exp. Sta. v. (for 1886), 150.

1887. Goff, E. S.   The Influence of Foreign Pollen upon the Character of the Fruit,—Strawberry and Grape.   5th Rep. N. Y. Exp. Sta. (for 1886) 179.

1887. Goldring, F.   The Flamingo plant.   (*Anthurium Scherzerianum.*) Am. Flor. ii. 466.

1887. H. E.   Cross-Breeding of Wheat.   Gar. World, iv. 6.

1887. Laxton, T.   The Tendency to Heredity or Reversion Exhibited by Plants, Especially Concerning Roses.   Gar. Chron. 3 ser. i. 387.

1887. Masters, M. T.   A New Bigeneric Hybrid, the Issue of *Phaius grandifolius* by *Calanthe Veitchii*.   Gar. Chron. 3 ser. i. 45.

1887. Masters, M. T.   Hybrid Lychnis.   Gar. Chron. 3 ser. ii. 56, 79, 99; Abstr. in Pop. Gar. iii. 31.

1887. Schiffner, V.   Ueber Verbascum Hybriden und einige neue Bastarde der *Verbascum pyramidalum*.   Bibliotheca Botanica, heft 3. 2 tab.

1887. Skeels, F. E.   Cross-Fertilization of Curcurbits.   Agric. Sci. i. 228.

1887. Sturtevant, E. L.   Influence of Crossing in Corn.   Rep. N. Y. Exp. Sta. v. (for 1886), 63.

1887. Thorpe. John.   Hybridizing and Cross-Fertilization.   Jour. Hort. v. 521, 569.

1887. W. H. W. C. New Potato Hybrids. Am. Gar. viii. 45.

1888. Bailey, L. H. Notes on Daturas—An Experiment in Hybridization. Bull. 40, Mich. Exp. Sta.

1888. Beverinck, M. Experiments in Hybridizing Barley. Acad. Sci. Holland ; Nature, Aug. 2, 1888 ; Abstract in Agric. Sci. ii. 243 (1888).

1888. Burbidge, F. W. The Tendency of Hybrids to Reproduce Themselves by Buds instead of by Seeds. Gar. Chron. 3 ser. iii. 179.

1888. Carman, E. S. Raspberry and Blackberry Hybrids. Rural N.-Y., Feb. 18. Note in Gar. and Forest, i. 372.

1888. Cassidy, James. Potatoes. Bull. 4, Colo. Exp. Sta.

1888. Crozier, A. A. Does the Effect of a Cross Appear in the Fruit of the First Year? Agric. Sci. ii. 319.

1888. Crozier, A. A. The Work in Crossing. [Immediate Influence of Pollen.] Bull. 3, Iowa Exp. Sta. 91-92.

1888. (Editorial.) Le Phylloxera et les parasites végétaux vaincus par l'hybridation. Rev. Hort. 50.

1888. Foster, M. Concerning the Supposed Tendency of Hybrids to Reproduce by Buds to a Greater Degree than by Seeds. Gar. Chron. 3 ser. iii. 151, 212.

1888. Hogg, James. Origin of the Le Conte Pear. Gar. and Forest, i. 392.

1888. Hooper, David. The Hybridization of Cinchonas. Tropical Agriculturist, viii. 6. Abstract in Agric. Sci. iii. 41 (1889).

1888. Hoopes, J. Hybrid Aquilegias. Gar. and Forest, i. 114.

1888. J. D. Experiments in Hybridization and Selection. Gar. Mag. xxxi. 592.

1888. Noble, Chas. The Dissociation of Hybrid Characters as Illustrated by *Clematis Jackmanni* var. *alba.* Gar. Chron. 3 ser. iv. 152.

1888. Tracy, W. W. Experiments in Crossing Corn, Tomatoes and Carrots. Rep. Mich. Hort. Soc. 43.

1888. Vilmorin, Henry L. de. Expériences de Croisement entre des Blés Différents, pp. 4. 2 tab. Paris.

1889. Bailey, L. H. Crossing of Squashes. Bull. 48. Mich. Exp. Sta. 8.

1889. Beckwith, M. H. Raspberry Hybrids. Rep. N. Y. Exp. Sta. vii. (for 1888), 233.

1889. Briggs, T. R. A. Hybrid Thistles. Jour. of Botany, xxvii. 270.

1889. "Calypso." Progress in Hybridizing Orchids. Gar. and Forest, ii. 367.

1889. Carman, E. S. Raspberry × Blackberry. Rural N.-Y. Nov. 23.

1889. Castle, Lewis. Hybrid Cactuses. Jour. Hort. xviii. 3 ser. 349.

1889. (Editorial.) The Finest Hybrid Orchid. Gar. Mag. xxxii. 286.

1889. (Editorial.) The Influence of Parentage upon Hybrids. Gar. Mag. xxxii. 732.

1889. Goodale, Geo. L. Hybridization. Gar. and Forest, ii. 188.

1889. Hays, W. M. Improving Corn by Cross-Fertilization and by Selection. Bull. 7, Minn. Exp. Sta. 27-33.

1889. Jackson, R. T. Hybridization of Gladioli. Gar. and Forest, ii. 88.

1889. J. H. J. Hybrid of Alpine Auricula with *Primula nivea*. Gar. xxxv. 375.

1889. Kellerman, W. A. and W. T. Swingle. Experiments in Cross-Fertilization of Corn. Rep. Kans. Exp. Sta. i. (for 1888), 316.

1889. Speer, R. P. Experiments with Corn. Bull. 7, Iowa Exp. Sta. 247-260.

1889. Wallace, Alfred Russell Chapters vi. and vii. in "Darwinism."

1890. Bailey, L. H. Experiences in Crossing Cucurbits. Bull. xxv. Cornell Exp. Sta. 180.

1890. Carman, E. S. Crossing and Hybridizing. Proc. Soc. Amer. Florists, 87 ; Rural N.-Y. xlix. (1890) 568 and 583 (Aug. 30 and Sept. 6); Am. Flor. vi. 109 ; Abstract in Gar. and Forest, iii. 421, and Pop. Gar. vi. 66.

1890. Crozier, A. A. Immediate Influence of Cross-Fertilization upon the Fruit.—Strawberries. Agric. Sci. iv. 287.

1890. Culverwell. Hybrid between the Gooseberry and the Black Currant. Gar. Chron. 3 ser. vii. 585.

1890. Dod, C. Wolley. Sterile and Fertile Hybrids of Hardy Plants. Gar. Chron. 3 ser. viii. 496.

1890. (Editorial.) Hybrid Firs. (Abstract of Paper by E. Bailly.) Gar. and Forest, iii. 308.

1890. Fryer, Alfred. A New Hybrid Potamogeton of the fluitans Group. Jour. of Botany, xxviii. 321.

1890. Hunn, C. E. Influence of Pollen upon Strawberries. Bull. 24, (n. s.) N. Y. State Exp. Sta. 329-330.

1890. Kellerman, W. A. and W. T. Swingle. Experiments in Crossing Varieties of Corn. Rep. Kans. Exp. Sta. ii. (for 1889), 288, 335.

1890. Kellerman, W. A. and W. T. Swingle. Crossed Varieties of Corn, Second and Third Years. Bull. 17, Kans. Exp. Sta.

1890. MacFarlane, J. The Microscopic Structure of Hybrids. Gar. Chron. 3 ser. vii. 543.

1890. Thorpe, John. Chrysanthemum Crosses. Gar. and Forest, iii. 546.

1890. Truffaut, Geo. Les Nephenthes ; leur fécondation. et leur hybridation. Hort. Belge, xvi. 78.

1890. Volxem, Jean Van. Hybridizing. Gar. Chron. 3 ser. vi. 141.

1890. Watson, W. Hybrid Hippeastrums. Gar. and Forest, iii. 165.

1891. Allen, Jas. Hybrid Between Chionodoxa and Scilla. Gar. xxxix. 308.

1891. Bailey, L. H. Crossing of Egg-Plants. Bull. 26, Cornell Exp. Sta. 14.

1891. —— Effect of Pollination upon Tomatoes. Bull. 28, Cornell Exp. Sta. 51.

1891. —— Pollination and Crossing of Cucumbers. Bull. 31, Cornell Exp. Sta. 134, 137.

1891. —— Tomato Crosses. Bull. 32, Cornell Exp. Sta. 165.

1891. Bessey, C. E. The Hybridization of Plants. Gar. and Forest, iv. 466. (Read before Am. Pom. Soc.)

1891. Budd, J. L. Breeding of the Orchard and Garden Fruits. Bull. 14, Iowa Exp. Sta. 181.

1891. Carman, E. S. Wheat-Rye Hybrids. Rural N.-Y. l. 653 (Sept. 12).

1891. —— Blackberry-Raspberry Hybrids. Rural N.-Y. l. 670 (Sept. 19).

1891. Douglas, J Cross-Bred Orchids. Gar. Chron. x. 396.

1891. Druery, Chas. T. Fern Hybrids and Crosses. Gar. Mag. No. 1949, p. 138.

1891. Em. R. Anthurium $\times$ rotundispathum. Ill. Hort. xxxviii. 5 ser. 9.

1891. Engleheart, G. H. Hybrid Narcissi. Gar. Chron. ix. 702; Gar. Mag. No. 1964, 367.

1891. Foussat, J. Les Glaïeuls hybrides de M. Lemoine. Rev. Hort. 32.

1891. Fraser, P. N. Crossing of Different Species. Gar. Chron. ix. 565.

1891. (Editorial.) A new Hybrid Rose. Gar. and Forest, iv. 532.

1891. Endicott, W. E. Some Hybrid Gladioli. Gar. and Forest, iv. 403.

1891. G. R. Hybrid Lilies. Gar. Mag. No. 1945, 86.

1891. L. L. *Vriesea cardinalis*, Hybrid. Ill. Hort. xxxviii. 39.

1891. McFarlane, J. M. Anatomical Structure of Hybrids. Jour. of Hort. xxiii. 86.

1891. McFarlane, J. M. The Color, Flowering Period and Constitutional Vigor of Hybrids. Gar. Chron. ix. 753.

1891. Meehan, Thos. On the Varying Character of Hybrids. Gar. Chron. x. 109.

1891. Oberswetter. Hybrid between a Crinum and Amaryllis. Jour. of Hort. x. 347.

1891. W. S. Hybrid Auriculas. Gar. Chron. ix. 3 ser. 557.

www.ingramcontent.com/pod-product-compliance
Lightning Source LLC
Chambersburg PA
CBHW022030190326
41519CB00010B/1647